Principles of Biotechnology

Principles of Biotechnology

Second Edition

edited by

ALAN WISEMAN, PhD, FRSC, MIBiol
Senior Lecturer, Department of Biochemistry
University of Surrey

Surrey University Press

Published in the USA by
Chapman and Hall
New York

Published by Surrey University Press
Bishopbriggs, Glasgow G64 2NZ and
7 Leicester Place, London WC 2H 7BP

Published in the USA by
Chapman and Hall
a division of Routledge, Chapman and Hall, Inc.
29 West 35th Street, New York, NY 10001-2291

© 1988 Blackie and Son Ltd.
First published 1983
Reprinted 1985, 1986
This edition 1988

*All rights reserved. No part of this publication
may be reproduced, stored in a retrieval system,
or transmitted, in any form or by any means,
electronic, mechanical, recording or otherwise,
without prior permission of the Publishers.*

British Library Cataloguing in Publication Data

Principles of biotechnology.——2nd ed.
  1. Biotechnology
  I. Wiseman, Alan, *1936–*
  660'.6

ISBN 0-903384-60-4

Library of Congress Cataloging-in-Publication Data

Principles of biotechnology/edited by Alan Wiseman.——2nd ed.
    p.   cm.
  Includes bibliographies and index.
  ISBN 0-412-01791-1 (Chapman and Hall: pbk.)
  1. Biotechnology.   I. Wiseman, Alan.
TP248.2.P753   1988
660'.6——dc19
                                              88-4963
                                                  CIP

Printed in Great Britain by Bell & Bain (Glasgow) Ltd

Preface

The first edition of this book appeared in 1983, and provided the first easily-accessible account of the state of biotechnology at a level suitable for advanced undergraduates and postgraduates. In this new edition, specialists in biotechnology at the University of Surrey have again collaborated with industrial experts to provide authoritative interdisciplinary coverage of an ever-expanding field.

The revision of the text reflects the rapid advances made in many of the fields which go to make up biotechnology—ranging from the molecular to the process plant level. Biotechnology is an applied science, and this book relates theory to application by placing the basic advances in the context of commercial applicability and process optimization. For those wishing to explore the literature further, references to up-to-date reviews and original research publications are provided.

Special thanks are due not only to the contributors, but to all those who have helped in the planning of this book.

AW

Background to authors

C. Bucke is Principal Lecturer in the School of Biotechnology, Polytechnic of Central London, and was previously Programme Manager Biotechnology at Tate & Lyle Group Research and Development. His research interests include the production and use of immobilized biocatalysts, the use of enzymes in the synthesis of novel bisaccharides and polysaccharides and the use of enzymes in extreme environments. He is Scientific Co-ordinator of the Institute for Biotechnological Studies, Department of Trade and Industry 'Extended use of Biocatalysts' Programme.

Mike Bushell is Lecturer in Industrial Microbiology in the University of Surrey. His research interests are in microbial process design, with particular reference to the production of secondary metabolites. Before joining the university, he spent five years in the pharmaceutical industry.

Peter Cheetham is Biosciences Manager at Quest International. While employed at Tate & Lyle Group Research and Development, Reading, he worked primarily on the enzyme and microbial production of novel sugars and sweeteners, especially using immobilized biocatalysts and applications to starch processing and high-intensity sweetener synthesis. He joined PPF International (now Quest International) in 1986, working on the applications of enzymology, fermentation and 'mild' chemistry to the production of food flavours and ingredients and fragrances, especially natural flavour chemicals.

Jeremy Dale is Senior Lecturer in Microbial Genetics and Medical Microbiology in the University of Surrey. His research interests are centred on the application of microbial genetics, including genetic manipulation techniques, to medically important bacteria, especially the causative agents of tuberculosis and leprosy.

Michael Winkler is Lecturer in Biochemical Engineering in the Department of Chemical and Process Engineering, University of Surrey. Before joining the university, he spent seven years in production management with a multinational food company, working in the UK, Scandinavia and West Africa. He is an adviser on biotechnology to the National University of Malaysia and the Malaysian University of Agriculture.

Alan Wiseman is Senior Lecturer in Biochemistry in the University of Surrey. He is editor or author of seventeen books and over one hundred papers on proteins and enzymes, and is a member of the editorial board of *Journal of Chemical Technology and Biotechnology* and of *Chemistry and Industry*.

Contents

1	**Features of biotechnology and its scientific basis** Alan Wiseman	**1**
1.1	Introduction	1
1.2	Interrelationships between microorganisms and enzymes	2
1.3	Success in biotechnology	2
2	**Application of the principles of industrial microbiology to biotechnology** M. E. Bushell	**5**
2.1	Introduction	5
2.2	Primary metabolism	5
	2.2.1 Introduction	5
	2.2.2 The batch culture	6
2.3	Secondary metabolism	7
	2.3.1 Products	7
	2.3.2 Definitions of a secondary metabolite	7
	2.3.3 Regulatory factors	8
2.4	Primary metabolites in industrial biotechnology	10
	2.4.1 Potable alcohol	10
	2.4.2 Amino acids	13
	2.4.3 Other organic acids	17
	2.4.4 Polysaccharides	19
	2.4.5 Other primary metabolites	21
	2.4.6 Single cell protein (SCP)	22
	2.4.7 Future plans in the UK	30
2.5	Secondary metabolites in biotechnology	30
	2.5.1 Penicillin	30
	2.5.2 Other secondary metabolites	38
3	**Application of the principles of microbial genetics to biotechnology** J. W. Dale	**44**
3.1	Control mechanisms in microorganisms	44

3.2	Manipulations *in vivo*	48
	3.2.1 Mutation	49
	3.2.2 Recombination	51
	3.2.3 Application of manipulations *in vivo* to increase enzyme production	54
	3.2.4 Overproduction of primary metabolites	55
	3.2.5 Overproduction of secondary metabolites	56
	3.2.6 Production of novel metabolites	61
3.3	Manipulations *in vitro*	61
	3.3.1 Basic techniques of *in-vitro* genetic manipulation	62
	3.3.2 Uses of DNA cloning	69
	3.3.3 Genes and biotechnology: applications of gene cloning	73
	3.3.4 Safety implications	78
	3.3.5 Future prospects	79

4   Application of the principles of fermentation engineering to biotechnology — 83
M. A. Winkler

4.1	The fermenter	86
	4.1.1 The environment	86
	4.1.2 Principal types of fermenter	87
	4.1.3 Mode of operation	90
	4.1.4 Ancillary processes	96
4.2	General design principles	97
	4.2.1 Basic design rules	98
	4.2.2 Materials and components	99
	4.2.3 Control and instrumentation	101
	4.2.4 Stirred-tank fermenters	104
	4.2.5 Gas-lift and sparged-tank fermenters	107
4.3	Heat transfer	110
	4.3.1 Heat sterilization	113
	4.3.2 Sterilization without heat	114
4.4	Mixing	116
	4.4.1 Introduction	116
	4.4.2 Outline of rheology	118
4.5	Oxygen supply in fermenters	122
4.6	Scale-up in biotechnology	126
	4.6.1 The problems of large-scale operation	126
	4.6.2 Selection of scale-up criteria	127
	4.6.3 Interaction of criteria	127

4.7	Fermentation processes in biotechnology	128
	4.7.1 Brewing	128
	4.7.2 Penicillin manufacture	129
	4.7.3 Biological waste-water treatment	130
	4.7.4 Single-cell protein (SCP) production	131
	4.7.5 Outlook for biotechnology	133
4.8	Summary	133

5 Application of the principles of enzymology to biotechnology — 136
Alan Wiseman

5.1	Features of enzymes in relation to biotechnology	136
	5.1.1 Introduction	136
	5.1.2 Advantages of using enzymes for manufacture of products	136
	5.1.3 Choice and control of enzymes in applications	138
5.2	Applications of enzymes in biotechnology	139
	5.2.1 Large-scale industrial applications	139

6 The biotechnology of enzyme isolation and purification — 143
C. Bucke

6.1	Introduction	143
6.2	Enzyme sources	143
6.3	Release of enzymes from cells	143
	6.3.1 Sources	144
	6.3.2 Extraction by physical methods	145
	6.3.3 Extraction by chemical methods	148
6.4	Primary clarification of the soluble enzyme	149
	6.4.1 Centrifugation	149
	6.4.2 Flocculation and coagulation	151
	6.4.3 Filtration	152
6.5	Concentration	153
	6.5.1 Removal of nucleic acids	153
	6.5.2 Precipitation	153
	6.5.3 Ultrafiltration and reverse osmosis	154
	6.5.4 Freeze-drying	156
	6.5.5 Evaporation	156
	6.5.6 Freezing	156
6.6	Enzyme purification — chromatography	157
	6.6.1 Gel chromatography	157
	6.6.2 Ion exchange chromatography	159
	6.6.3 Affinity purification	161

	6.6.4	Chromatography columns	162
	6.6.5	High-performance liquid chromatography of enzymes	163

7 The application of immobilized enzymes, immobilized cells and biochemical reactors in biotechnology—principles of enzyme engineering
Peter S. J. Cheetham
  164

7.1	Introduction		164
7.2	The application of biological catalysts		164
7.3	Types of enzymic catalyst and commercial applications		165
	7.3.1	Immobilized biocatalysts	168
	7.3.2	Assessment of supports and methods	173
	7.3.3	Effectiveness factors for immobilized enzymes	174
	7.3.4	The kinetics of enzymes in industrial use	174
	7.3.5	Factors which modify the intrinsic activity of enzymes	177
	7.3.6	The stability of immobilized biocatalyst—diffusion artefacts	180
7.4	Enzyme reactors		181
	7.4.1	Types of biochemical reactor	183
	7.4.2	Assessment of the performance of biochemical reactors	183
	7.4.3	Practical enzyme reactor kinetics	190
	7.4.4	The effect of non-ideal flow on biochemical reactor performance	192
	7.4.5	The stability of biochemical reactors	193
	7.4.6	Physical problems associated with the use of immobilized biocatalysts in biochemical reactors	196
	7.4.7	Purification and recovery of the products of biochemical reactors (downstream processing)	197
7.5	Conclusions		200

Index   203

1 Features of biotechnology and its scientific basis
ALAN WISEMAN

1.1 Introduction

The Spinks Report (1980) defined biotechnology as *the application of biological organisms, systems or processes to manufacturing and service industries*. It was envisaged that biotechnology would create new industries, with low energy demands. This is because the growth of microorganisms provides us with a renewable source of energy, so that it is not necessary to rely on the diminishing stocks worldwide of expensive fossil-fuels and chemicals derived from them. Microorganisms have many uses, including the production of chemicals such as ethanol by fermentation procedures, brewing with yeast, production of enzymes and as a foodstuff (single cell protein, SCP). Genetically-engineered microorganisms can be used to manufacture human proteins and enzymes for medical application.

It is clear that an appreciation of the potential of biotechnology depends on a thorough understanding of the basic sciences involved. Nevertheless it is easy to identify the areas where natural raw materials, including foodstuffs and agricultural wastes, can be converted catalytically by microorganisms, or their derived enzymes, to useful products. Such processes include traditional ones such as brewing, which is a part of enzyme and fermentation biotechnology. Other areas include waste treatment, to avoid environmental pollution, and here the organisms grown (biomass, SCP) can be used as animal food or as a source of chemicals such as methane. Much of biotechnology involves the *discovery* and subsequent *optimization* of the biological and biochemical processes needed to exploit the source of natural raw material.

Teaching and learning in a field as diverse as biotechnology is hampered by the inability of students and specialists in each of the constituent scientific areas—microbiology, chemical engineering, biochemistry and chemistry—to understand each other's language. This is mainly because each lacks the basic knowledge required to understand and interpret even the most fundamental of the concepts under discussion. The biochemist needs to accept that some knowledge and understanding of the principles of chemical engineering is essential to the biotechnologist/biochemist. Similarly the biotechnologist/ chemical engineer must understand something about the biochemistry of enzymes and the microbiology of living organisms in the reactor (for specialist review series see Wiseman, 1985, 1977–1985). It usually proves difficult to cross interdisciplinary boundaries in the search for understanding (and so professional competence) in one's own area of biotechnological work. Even the questions we would like to ask a subject specialist cannot easily be formulated because of differences in language, jargon and definitions. This is as true for the young student as for the professionally qualified specialist in any one of the three main areas covered in this book.

A preliminary idea of what biotechnology is, as expressed through the industrial use of the basic sciences underpinning it, is provided in Smith's excellent introductory outline (1981). Details of these basic sciences in relation to their potential application are covered in the present book, which aims to provide a fundamental understanding of the advantages and problems inherent in their application. We hope that our book will allow a start to be made in the formulation of questions and answers in the general field of biotechnology.

1.2 Interrelationships between microorganisms and enzymes

The relationship between living microorganisms and the enzymes they contain becomes important when we wish to exploit the enzymes to perform specific chemical conversions. The enzyme can be isolated (Chapter 6), and enzyme biotechnology is concerned with the efficient handling of the isolated enzyme in appropriate reactors (Chapter 7). However, microbial cells, especially when immobilized on solid supports, act as catalysts for particular chemical conversions, much like the isolated enzymes themselves, giving the biotechnologist less preparative work. Much of fermentation technology deals with growing microorganisms on a large scale (Chapter 4).

Selection and genetic engineering of appropriate microorganisms (Chapter 3) allows the enzyme content to be manipulated in favourable cases. Also, growth on a particular substrate will often cause the induction of the enzyme needed for the breakdown of that growth substrate within the cell. For example, growing yeast on maltose will induce biosynthesis of the appropriate enzyme, maltase (α-glucosidase) which breaks down the maltose to the glucose needed for production of chemical energy in the form of adenosine triphosphate (ATP) within the yeast cell. Similarly, in *Escherichia coli*, lactose and other inducers cause the production of lactase (β-galactosidase) by switch-on of the *lac* operon DNA (Chapter 3).

1.3 Success in biotechnology

The interdisciplinary nature of biotechnology is a direct result of interrelationships between apparently disparate phenomena and considerations. The need to bring together chemical engineers, microbiologists, biochemists and chemists arises from the essential requirement for success in the *practical* undertakings of biotechnology.

Success in biotechnology means economic success as well as scientific success. It is essential therefore that process engineers are involved at an early stage in the planning of a venture in biotechnology. Often it is not clear at this stage if the biological, biochemical and chemical understanding of the required system is sufficient to achieve the scientific success. The further requirement for economic success adds an additional critical appraisal of any intended application, especially on a large scale or where considerable investment is necessary in materials, fuel and manpower. Planning consideration may often favour the use of continuous processes rather than batch processes for these reasons. The advantages and disadvantages, and therefore the choice of batchwise or continuous methods of processing or production are not always obvious, and may be controversial. Further discussion, for example involving the production of beer, has been documented by Kirsop (1982).

Much further research is proceeding in universities, in research institutes and in industry. In the UK, frequent contact and communication is maintained through the British Co-ordinating Committee on Biotechnology (BCCB) with the support of the Society of Chemical Industry (and its Biotechnology Group) and other learned socities, while in Europe, the European Federation of Biotechnology (EFB) is active in promoting such contact and communication.

Worldwide interest is unmistakable, as seen through the frequent coverage of biotechnology in the media and in journals and books. Useful discoveries are inevitable by use of techniques of genetic engineering to put the required genes (human genes on occasion) into the appropriate microorganism, to make the product that is really needed by the market. Human proteins and enzymes required for therapy are being made in this way. Many other benefits will result from the manufacture, by fermentation techniques, of specific antibodies for the purposes of analysis and therapy. These monoclonal antibodies are produced by the growth of cells in large culture vessels, using the prior biotechnological knowledge and understanding gained by growing microorganisms in large fermenters—for example, for the production of antibiotics such as penicillins. New developments and commercial applications are appearing rapidly in every field of biotechnology, including the fermentation industries, enzyme and immobilized cell biotechnology, waste treatment and by-product utilization. The successful processes will be useful to society, attractive to industries for commercial reasons and supported by Governments where appropriate.

References

Kirsop, B. (1982) *Topics in Enzyme and Fermentation Biotechnology*, ed. A. Wiseman, vol. 6, Ellis Horwood, Chichester.

Smith, J. E. (1981) *Biotechnology*. The Institute of Biology, Studies in Biology No. 136, Edward Arnold, London, 1–73.

Spinks, A., *Biotechnology*, report of a Joint Working Party, HMSO London, March 1980. (Joint Working Party—Advisory Council for Applied Research and Development, Advisory Board for the Research Councils, and The Royal Society).

Wiseman, A. (1985) *Handbook of Enzyme Biotechnology*, ed. A. Wiseman, 2nd edn., Ellis Horwood, Chichester.

Wiseman, A. (1977–1985), ed., *Topics in Enzyme and Fermentation Biotechnology*, vols. 1–10, Ellis Horwood, Chichester.

2 Application of the principles of industrial microbiology to biotechnology
M. E. BUSHELL

2.1 Introduction

In process design, one function of the industrial microbiologist is to investigate the important aspects of the physiology of the producing organism which control the extent and circumstances of product synthesis. Such information is important in evolving a working process control model and in optimization studies during a yield improvement programme.

Fortunately it is possible to categorize, in many cases, the dynamics of product formation on the basics of past experience and make predictions which are useful during subsequent process design. As this chapter is intended to provide an introduction to microbial processes, a rather broad division has been made, classifying natural products as primary or secondary metabolites. This exercise is only useful in recognizing the basic characteristics of newly discovered products so that rapid process optimization may be initiated, drawing on past experience of similar types of product. It should not be allowed to act as a conceptual strait-jacket, forcing the reader to consider all products as either primary or secondary metabolites. Indeed, the more information one obtains on an individual product, the more difficult it becomes to classify it as belonging to one particular class.

2.2 Primary metabolism

2.2.1 Introduction
When a microorganism grows in an environment in which all its essential nutrients are contained in excess, it converts those nutrients into compounds representing the end products of energy metabolism and compounds required to reproduce, thus synthesizing more cells. These two categories of compounds required for growth are known as *primary metabolites*. Commercially important primary metabolites include alcohol and other industrial solvents, amino acids and other organic acids, nucleotides, polysaccharides, fats, vitamins and enzymes. Also included in this category is cell biomass, perhaps the most obvious product of primary metabolism.

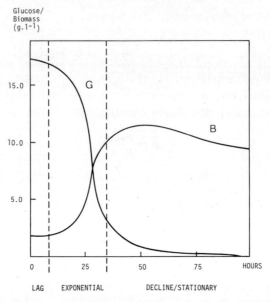

Figure 2.1 Biomass (B) accretion and glucose (G) consumption curves for *Streptomyces cattleya* in a batch culture.

2.2.2 *The batch culture* (see also 4.1.3)

When a microorganism is inoculated into a nutrient-rich environment, the rate of cell division, and therefore biomass production, varies according to a characteristic sequence (Figure 2.1). Phase 1, the lag phase, can often be virtually eliminated by increasing the size of the inoculum and optimizing the physical conditions within the fermenter. In a commercial process, the duration of the fermentation contributes to the cost of the product and a minimal lag phase is highly desirable. Phase 2 is known as the exponential phase, during which the cell division or biomass accretion rate becomes maximal. The rate of biomass production approximates to that represented by the equation

$$x_1 = x_0 e^{\mu t} \qquad (2.1)$$

where the biomass has increased from $x_0 \mathrm{g\,l^{-1}}$ in t hours. The constant μ is known as the specific growth rate of the organism (units, $\mathrm{h^{-1}}$) and is derived from the expression

$$\mu = \frac{dx}{dt} \cdot \frac{1}{x} \qquad (2.2)$$

where dx/dt is the rate of biomass increase (units, $\mathrm{g\,l^{-1}h^{-1}}$) at the point where biomass has reached x grams.

Integration and rearrangement of (2.2) gives

$$\ln x_1 = \ln x_0 + \mu t$$

The value of μ is often calculated by plotting $\ln x$ against t which, when μ is constant, will give a straight line of slope $\mu(\mathrm{h}^{-1})$. Careful measurement of constant, biomass, however, reveals that μ is never constant in a batch culture and varies according to the Monod equation

$$\mu = \frac{\mu_{max} s}{K_s + s} \quad (2.3)$$

where μ_{max} is the maximum specific growth rate observed in the batch culture, s is the concentration of growth-limiting nutrient in the culture medium and K_s is a constant whose magnitude represents the affinity of the organism for the growth-limiting substrate. As the growth-limiting substrate is consumed by the culture, its concentration s in the medium decreases until it becomes commensurate with K_s. From this point, the culture growth rate declines significantly (eqn 2.3) until a stationary phase (Figure 2.1) is reached, during which growth ceases altogether. A decline phase corresponding to cell lysis usually follows the stationary phase.

As the value of s approaches that of K_s, some microbial cultures synthesize, and often excrete, compounds known as secondary metabolites, substances which have provided the bulk of the commercial reward due to industrial microbial processes involving natural products.

2.3 Secondary metabolism

2.3.1 *Products*
Commercially, the most important secondary metabolites are the antibiotics. Of these, penicillin is produced in the greatest quantity. Other important antibiotics produced by fermentation are the aminoglycosides, cephalosporins, polyenes, non-polyene macrolides and the tetracyclines. Toxins and alkaloids account for the remainder of the secondary metabolites of industrial importance. More recently, attempts to isolate secondary metabolites with agrochemical potential have been made.

2.3.2 *Definitions of a secondary metabolite*
2.3.2.1 *Temporal.* In the previous section it was stated that secondary metabolites are produced when growth-limiting substrate concentration approaches the K_s value for the culture. This is simply an observation of the coincidence of two events and the exact physiological relationship between them has yet to be established.

2.3.2.2 *Physiology of the producing organism.* It has often been stated in the

Table 2.1 Secondary metabolites which may influence the physiology of the producing organism. (Information first compiled by Zahner, 1977.)

Metabolite	Function	Possible role
Sideramines (e.g. ferrichromes and ferrioxamines)	Iron, cobalt and nickel ion chelators	Scavengers in iron-, cobalt- or nickel-depleted environments.
Ionophore antibiotics (e.g. macrotetrolides)	Mobilize alkali metal ions	Increase membrane permeability to alkali metal ions selectively.
Antibiotics involved in spore formation (gramicidins)	Inhibit binding of RNA polymerase to DNA	Selectively inhibit the genes of vegetative metabolism in favour of those which promote sporulation.

literature that secondary metabolites have no recognizable function in physiology of the producing organism. The antibiotic function, for example, is usually derived from compounds excreted too late in the producing organism's life cycle to confer any ecological advantage. There are, however, a few secondary metabolites which may have some physiological significance to the producing organism (Table 2.1). The sideramines are part of a large group of compounds known as hydroxamic acids, and ferrichrome and ferrixamine hydroxamic acids have been shown to act as intermediates in iron assimilation into the cell. A similar function in alkali metal ion uptake has been ascribed to the ionophores. The sporulation antibiotics formed in *Bacillus subtilis* have been implicated in spore formation, but evidence for this requires substantiation.

A slight modification of Demain's view, 'secondary metabolites are not essential for growth', results in a good working definition: secondary metabolites are compounds which are not essential for exponential growth.

2.3.3 *Regulatory factors* (see also 3.1)

2.3.3.1 *Chromosomal involvement*. It has been established that many peptide antibiotics are synthesized in the microorganism using an alternative mechanism to its normal protein synthesis. Towards the end of 'exponential phase', enzymes called 'antibiotic-committed enzymes' have been detected in many product-forming microorganisms. Biosynthetic precursors of peptide antibiotics, such as glutamic acid, are activated by ATP-committed enzymes. An example of such a process is the synthesis of gramicidin S in *Bacillus brevis*.

The antibiotic-committed enzyme gramicidin synthetase II initiates the biosynthetic sequence by catalysing the activation of L-phenylalanine by ATP to form the aminoacyl derivative (Figure 2.2). This process is analogous to the formation of aminoacyl transfer RNA in primary metabolism where amino acids are activated prior to stepwise biosynthesis of protein molecules.

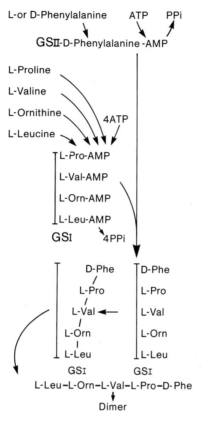

Figure 2.2 Biosynthesis of gramicidin S in *Bacillus brevis*.

Gramicidin synthetase II also effects the conversion of L-phenylalanine to the D-form which becomes bound to gramicidin synthetase I. This latter antibiotic-committed enzyme catalyses the adenylation of the other precursor amino acids (Figure 2.2) which then bind on to gramicidin synthetase I as thiol esters. The polymerization of all the amino acids results in the formation of linear gramicidin (a linear pentapeptide) which is then released from the enzyme complex. Linear gramicidin molecules then combine as dimers to form the final cyclic product, gramicidin S (a cyclic decapeptide).

Such detailed biosynthetic information is not available for the majority of secondary metabolites, but it is thought that the use of protein templates instead of messenger RNA may be a common (the 'thiotemplate' mechanism) feature of peptide antibiotics synthesis.

2.3.3.2 *Energy charge*. It has been proposed that energy charge, referring to the concentration of adenosine mono-, di- and triphosphate in the cells, may control antibiotic synthesis, where

$$\text{Energy charge} = \frac{[\text{ATP}] + 0.5\,[\text{ADP}]}{[\text{ATP}] + [\text{ADP}] + [\text{AMP}]}$$

This has been suggested because, for many antibiotics, culture phosphate concentrations which are optimal for exponential growth suppress antibiotic formation. When culture phosphate concentration drops to the prerequisite level, antibiotic synthesis takes place. The rate of phosphate flux into *Escherichia coli* cells has been shown to influence energy charge directly. A high flux increases ATP formation and so raises the charge level.

2.3.3.3 *Carbon catabolite repression.* Glucose, an excellent carbon and energy source for microbial growth, often exerts a repressive effect on antibiotics (e.g. tetracyclines and bactiracin) and the alkaloids. Glucose has been shown to inhibit the synthesis of antibiotic-committed enzymes, for example phenoxazinone synthetase. This enzyme is involved in the production of the antibiotic actinomycin by *Streptomyces antibioticus*.

There may exist in secondary metabolism a control mechanism similar to the catabolite repression process observed in the *Escherichia coli lac* operon system. Such a mechanism would appear to control the synthesis of antibiotic-committed enzymes.

2.3.3.4 *Summary of the essential features of secondary metabolism.* Many antibiotics appear to be assembled using an alternative mechanism to the normal ribosomal system. Analogues to t- and m-RNA exist in the form of antibiotic-committed enzymes. The control parameters for antibiotic-committed enzyme formation may provide the basis for regulatory factors for secondary metabolite formation and include induction, carbon catabolite repression and energy charge regulation. Induction may result from the addition or the accumulation of a metabolic intermediate.

2.4 Primary metabolites in industrial microbiology

2.4.1 *Potable alcohol*

2.4.1.1 *Substrates.* The most widely-produced primary metabolite is ethanol for human consumption. In most regions of the world, alcoholic beverages can be produced from locally-available substrates (Table 2.2). Raw materials range from lower molecular weight sugars (in fruit, honey or vegetable material) to starches (root or grain).

Wines are prepared primarily, sometimes exclusively, from the juice of fresh grapes. Red wines are produced from red grapes, utilizing juice, pulp, skins, seeds and sometimes stalks. White wines may be produced from either red or white grapes. The juice alone provides the fermentation substrate, and tannin and other pigment-containing components of the crush are avoided.

Sparkling wines, such as champagne, are manufactured from white wines to which sucrose and tannin are added.

Ales, beer and lager are made by mashing barley malt and other cereals with water to form a saccharide-rich substrate, known as wort. Barley malt is produced from sprouted barley grain, which contains a high proportion of

Table 2.2 Fermentation substrates for potable alcohol production (non-distilled). (Courtesy of J. E. Smith, Department of Microbiology, University of Surrey.)

Substrates	Beverage	Country	Saccharifying agent
Starchy			
(Barley and other cereals)	Ale	Belgium W Germany Canada Australia	Barley malt
	Lager	Worldwide (industrial countries)	Barley malt
Barley, rye, rice beet	Kvass	USSR	Barley and rye malt
Millet	Busa	USSR (Crimea)	
	Braga	Romania	
	Thumba	India	
Rice	Arak[1]	India, SE Asia	
	Busa	Turkestan, USSR	
	Pachwai	India	*Mucor* sp.
	Sake	Japan	*Aspergillus oryzae*
	Sonti	India	*Rhizopus* sp.
Rice (red)[2]	Anchu	Taiwan	
	Hung-chu	China	
Sorghum	Kaffir beer	Malawi	Sorghum malt *Aspergillus* sp. *Mucor rouxii*
	Merissa	Sudan	*Bacillus* spp.
Sweet potato	Awamori	Japan	
Sugary			
Agave spp. (sap)	Pulque	Mexico	
Apple (juice)	Cider	UK, France N America	
Grape (juice)	Wine	Temperate: N and S Hemisphere	(not required)
Honey	Mead	UK	
Pear (juice)	Perry	UK, France	
Palmyra (juice)	Toddy[3]	India, SE Asia	
Palm flower-stalk (juice)	Tuwak	Indonesia	

[1] With molasses, palm juice, chiefly distilled.
[2] Rice fermented with the pigmented *Monascus purpureus*.
[3] Or distilled as arak.

amylolytic and proteolytic enzymes. The grain is coarsely ground and mixed with rice, amize or wheat (starching cereals) to form a 'grist'. Mashing consists of making a warm aqueous suspension of grist and allowing the amylolytic and proteolytic enzymes from the barley to release low molecular weight saccharides from the cereal material. The precise details of mashing vary according to the final product required. The 'sweet wort' produced by mashing is then treated with hops or hop extract in a 'boiling' stage. During this process, flavouring components are extracted from the hops and a 'protein break' occurs during which a precipitate of protein coagulated by heat and hop tannins appears in the wort, an essential product clarification step. The brewing process is described in 4.7.1.

Distilled beverage production involves mashing, saccharifying (in some products), fermenting and distilling phases. Raw materials include a number of cereal grains, sugar cane products and other plant materials (Table 2.3). During the mashing process, malted and whole cereal grains are milled and processed using similar processes to those employed for beer production. The saccharification step allows the activity of α-amylase enzymes on the gelatinized starch in the mash. In American rye whisky production, this may be accomplished by the addition of commercial enzyme preparations.

2.4.1.2 *Fermenting organisms.* The majority of alcoholic fermentations are carried out by the yeast *Saccharomyces cerevisiae* or a related species. *S. cerevisiae* is grown in surface culture in ale production. Lager production employs *S. carlsbergensis*, which sinks to the bottom of the fermenter. *S. cerevisiae* and *S. carlsbergensis* are respectively termed top and bottom fermenting yeasts. *S. cerevisiae* var. *ellipsoideus* (*S. ellipsoideus*) strains are used in wine production, and other fermented products employ specific species (sake is produced from *S. saki*, etc). Many 'traditional' fermentations

Table 2.3 Raw materials fermented to produce distilled spiritous liquors. After Ayres, J. C., *et al.* (1980) *Microbiology of Foods* (Freeman, San Francisco).

Liquor	Material fermented
Scotch whisky	Barley
Irish whiskey	Malt, unmalted barley, wheat, rye and oats
Bourbon	Corn
Canadian whiskey	Rye
Gin	Grain mash, flavoured with juniper berries, anise, etc.
Rum	Sugar cane
Aquavit	Grain or potatoes, flavoured with caraway seeds
Vodka	Grain or potatoes, not flavoured
Tequila (Mescal)	Juice from the hearts of the cactus *Agave tequilana*
Brandy	Various fruits
Kirschwasser	Cherry juice
Applejack	Apple juice
Cognac	White grapes produced in the Cognac region of France

employ mixed cultures of yeast and other microorganisms which form nutritionally self-contained communities (see section 2.4.6.5) which are resistant to further microbial contamination.

Most alcohol-producing yeasts are capable of hydrolysing disaccharides and many can split higher molecular weight sugars. All of them are able to assimilate hexoses, such as glucose and fructose, utilizing the Embden–Myerhof–Parnas (glycosis) pathway of fermentation for anaerobic ethanol production.

Fortified wines are produced by adding distilled spirits to wines. The initial fermentation for sherry production is a two-stage process. Wine is incubated under aerobic conditions with pellicle-forming yeasts (*S. beticus*) which converts ethanol and glycerol into aldehydes and other flavouring compounds.

The technology and processing parameters for alcoholic beverage fermentations are more fully described in Chapter 4.

2.4.2 *Amino acids*

Many amino acids are currently produced in commercial quantities through fermentation processes. They are used to supplement animal and human foodstuffs and have numerous medical applications, particularly in diet deficiency therapy (Table 2.4). Most of the world's amino acid fermentation capacity is located in Japan, where the annual production rate is twice that of the rest of the world.

Table 2.4 Amino acids produced in submerged culture

	Annual production (tons)	Use
L-arginine	250	Alleviation of hyper-ammonemia and liver disorders by stimulating arginase Stimulates insulin secretion
L-citrulline	30	Similar to arginine
L-glutamate	100 000	Food seasoning
L-histidine	150	B_6 deficiency and in pregnancy
L-homoserine	30	Deficiency diseases
L-glutamine	250	As histidine, and treatment of gastric ulcers
L-isoleucine	30	Deficiency diseases
L-leucine	75	Deficiency diseases
L-lysine	10 000	Animal feed additive, increases nutritive value of rice and cereal products
L-ornithine	30	Similar to arginine
L-phenylalanine	75	Deficiency diseases
L-proline	30	Deficiency diseases
L-threonine	75	Dietary uses
L-valine	75	Deficiency diseases
L-cysteine	Under development	Bronchitis treatment, post-operative intravenous infusion

2.4.2.1 *Producing organisms.* Commercial fermentations for *L-glutamic acid* have been described in the patent literature. They make use of a number of Gram positive, biotin-requiring, non-motile bacteria of the genera *Arthrobacter*, *Brevibacterium* and *Corynebacterium*. These genera are closely linked taxonomically (DNA guanosine–cytosine content 51.2–54.4 mole %) and are sometimes known as the glutamic-acid bacteria. Growing these biotin-requiring bacteria in biotin-limited media results in a decrease in cell membrane phospholipid content. The resultant decrease in membrane permeability facilitates glutamic acid secretion. Permeability may be further increased by the addition of penicillin and surfactants to the medium, sometimes to the extent that production can take place in the presence of excess biotin. Patents for *L-glutamine* production specify that glutamic acid bacteria are cultivated at pH values below 7 in media containing a high concentration of ammonium chloride and zinc ions. The interconversion between glutamate and glutamine is subject to so many types of metabolic control, however, that further experimentation will be required before the significance of these culture conditions can be assessed.

Production of *L-proline* and *L-lysine* is accomplished using auxotrophic

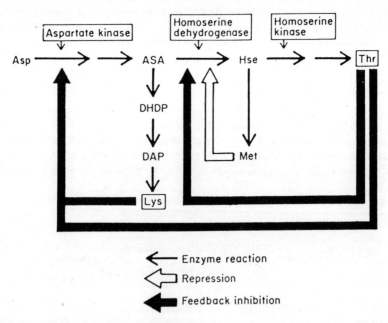

Figure 2.3 Regulation in lysine biosynthesis in *Brevibacterium flavum* and *Corynebacterium glutamicum*. ASA, aspartate semialdehyde; DHDP, dihydrodipicolinate; Hse, homoserine; DAP, diaminopimelate. Repression stops enzyme biosynthesis, while feedback inhibition stops the activity of an enzyme. From Hirose and Okada (1979), with permission.

bacteria. L-proline production employs an isoleucine-requiring auxotroph of *Brevibacterium flavum*. The metabolic link between proline and isoleucine is not immediately apparent. Isoleucine is one of the family of amino acids which in the bacterium can be derived from pyruvate, whereas proline is derived from glutamate. It has been suggested that isoleucine auxotrophy allows the accumulation of threonine, which results in feedback inhibition of aspartate kinase (Figure 2.3). Aspartate kinase has a higher affinity for ATP than that of glutamate kinase (the rate-limiting enzyme step in proline synthesis). Thus, inhibition of aspartate kinase would allow glutamate kinase to compete more successfully for pool ATP, which it requires for activity.

The first industrial use of auxotrophic mutants in bulk amino acid synthesis took place in L-lysine production. Industrial strains are either homoserine auxotrophs or methionine-serine double auxotrophs (Hirose and Okada, 1979). Strains thus lacking the ability to form homoserine or methionine and threonine will favour the lysine biosynthetic pathway branch (Figure 2.3) at aspartate semialdehyde. The rate-limiting step in lysine synthesis in such auxotrophic mutants results from intracellular accumulation of lysine. This gives rise to feedback inhibition at aspartate kinase (Figure 2.3). In an effort to overcome this limitation, screens have been mounted for mutants which have an aspartate kinase with an inactivated regulatory site. The enzyme is, however, subject to feedback inhibition from threonine as well as lysine. In order to avoid the accumulation of both amino acids, the double auxotroph (methionine–threonine) may be used as the parent of the deregulated strain. Thus, organisms combining auxotrophy and enzyme deregulation are now available for commercial use for L-lysine production.

A similar strategy has been applied to the selection of deregulated mutants for *L-phenylalanine* production. Strains with a prephenate hydratase (Figure 2.4) which is resistant to L-phenylalanine inhibition, have been developed for production processes. Mutants with deregulated enzymes are selected by screening for mutated strains which are able to grow in the presence of the

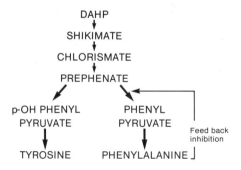

Figure 2.4 Feedback regulation in L-phenylalanine synthesis.

appropriate amino acid analogues. Such analogues bind to the regulatory site of the non-deregulated mutants, cutting off amino acid biosynthesis, thereby causing cell death. Deregulated mutants, with ineffective binding sites, survive this screening procedure.

2.4.2.2 *Fermentation principles in amino acid production.* Amino acids are produced in conventional stirred tank reactors (see 2.5). Fermenter and process designs share many characteristics with antibiotic production methods. Some differences arise in that most amino acid fermentations employ unicells whereas antibiotic processes are often based on filamentous organisms.

The commercial production of *L-isoleucine* is accomplished by the addition of α-aminobutyrate to the culture medium. This compound affects the fermentation in four ways.

(i) The α-aminobutyrate acts as a precursor to α-oxobutyrate (Figure 2.5), an intermediate in isoleucine biosynthesis
(ii) The compound acts as an enzyme deregulator, releasing the feedback inhibition of threonine dehydratase (Figure 2.5) by isoleucine
(iii) The α-aminobutyrate affects the enzyme, acetohydroxyacid synthetase, which catalyses the reversible interconversion between α-oxobutyrate (isoleucine precursor) and α-acetolactate (valine precursor); the presence of α-aminobutyrate favours the synthesis of the former
(iv) Finally, α-aminobutyrate acts as an amino donor in the transamination step between ketomethylvalerate and isoleucine

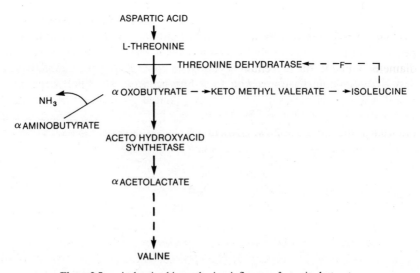

Figure 2.5 L-isoleucine biosynthesis—influence of α-aminobutyrate.

Other examples of precursor addition include the production of L-tryptophan by *Hansenula anomala* growing on glucose, which has been improved by the addition of anthranilic acid (a tryptophan precursor) to the medium. Selection of strains capable of maximizing this substrate fermentation has resulted in a process with a 90% yield at $3\,g\,l^{-1}$ tryptophan.

Similarly, L-serine is produced by *Corynebacterim glycinophilum* growing on a glycine/glucose mixture. Approximately $20\,g\,l^{-1}$ of glycine is converted into $10\,g\,l^{-1}$ of L-serine per $100\,g\,l^{-1}$ of glucose.

2.4.3 Other organic acids

2.4.3.1 Citric acid.
Citric acid is sold in the anhydrous and monohydrated forms and is the most widely used acidulant in the food industry. Other uses include descaling and metal cleaning. Sodium citrate is used in processed cheese manufacture and as a blood anticoagulant.

Commercial production is from *Aspergillus niger*, in which the Embden–Meyerhof–Parnas (glycolysis) pathway is responsible for the conversion of hexoses to pyruvate. This is converted via the tricarboxylic acid (Krebs) cycle into citrate.

Aconitase, which converts citric acid to *cis*-aconite in the cycle, requires iron for its activity. The production medium, therefore, is formulated with the minimum iron concentration necessary for growth without nutrient limitation, and contains copper, which acts as an antagonist of iron in the aconitase reaction.

In order to produce the citric acid at maximum yield, the culture must first assume the pelleted morphological form. Many fungi and actinomycetes, when growing in the mycelial form, do not form a uniform suspension of hyphae. Instead, they grow as aggregates, up to 10 mm across, known as pellets. The physiology of the mycelium within the pellet varies across the diameter. Outside, the mycelium is young, i.e. capable of exponential growth, and is physiologically equivalent to a homogeneous mycelial suspension. Inside, the mycelium is subject to oxygen-limited growth and may contain vacuolated or lysed hyphae. Electron micrographs of cross-sections of pellets often show them to be hollow. Growth of pellets is not exponential but linear, and proceeds according to a cube root law:

$$x_t^{1/3} = x_0^{1/3} + kt$$

Where $x_t\,g\,l^{-1}$ of biomass is produced from $x_0\,g\,l^{-1}$ after t hours, and k is a constant.

The oxygen limitation apparent in pelleted *Aspergillus niger* cultures may stimulate citric acid production in the organism. For this reason, production is carried out in airlift fermenters (see Chapter 4) which favour the pelleted form.

The principal carbon source employed is molasses, a by-product of the sugar-refining industry. A citric yield of 70%, on a carbon basis, is obtained. Molasses-based media take 7 days to achieve the required cell density, which is followed by a further 2 days for complete differentiation into pellets. Empirical experimentation has shown that a culture pH of 3.5 or less decreases the tendency to form oxalic or gluconic acids. Restricting the supply of nitrogen to the culture has a similar effect in discouraging oxalic acid formation.

Japanese workers have reported the production of citric and isocitric acids by *Candida lipolytica* growing on *n*-paraffins. The resultant process was faster and higher-yielding than the molasses-based fermentation. Since the development of this process, however, alterations in petroleum refining practice and price increases have resulted in *n*-paraffins becoming a less attractive substrate.

2.4.3.2 *Itaconic acid.* Itaconic acid is used in the production of plastics as, together with its esters, it is readily polymerized. It is also copolymerized with acrylic acid, methyl acrylate and styrene.

Itaconic acid is produced from *Aspergillus terreus*. A branched metabolic pathway from the TCA (Krebs) cycle intermediate *cis*-aconitate is involved, one branch leading to itaconic acid production. Fermentation media therefore include a high calcium ion concentration. Free icatonic acid is toxic to moulds at concentrations in excess of 7% but, with stepwise addition of ammonia to neutralize the acid, concentrations of up to 20% may be accumulated in the medium.

2.4.3.3 *Gluconic acid.* Calcium gluconate is a widely prescribed therapeutic agent in cases of calcium deficiency. Ferrous gluconate is, similarly, the preferred compound for iron administration in deficiency therapy. The free acid is employed as a mild acidulant in metal processing, leather tanning and the food industry. Glucose oxidase, an enzyme from *Aspergillus niger* which is also produced in the gluconic acid production process, is used for the stabilization of colour and flavour in beer, tinned foods and soft drinks. Glucose oxidase is used as a basis for the production of diagnostic kits to assay glucose in urine, and so is valuable in the evaluation and control of diabetes.

Gluconic acid is produced commercially in *Aspergillus niger* cultures. The growth medium is based on glucose and corn steep liquor. The process must be highly aerobic, and efficient oxygenation is ensured by the use of turbine impellers in baffled culture vessels. The gluconic acid is neutralized as it is formed by addition of sodium hydroxide solution.

2.4.3.4 *Acetic acid in vinegar.* Two types of vinegar are sold in the UK, vinegar derived from wine souring and that derived from beer (malt vinegar). As with many traditional fermented products, strict legal definitions limit the composition of the raw materials which may be used as substrates. Wine vinegar, for example, may only be produced by 'acetous fermentation of natural wine', and wine residues may not be used. Malt vinegar may only bear the name if it is produced from malted barley, with or without the addition of cereal grain, the starch of which has been converted only by the 'diastase' enzyme activity (amylase) of the malted barley.

Vinegar production is a two-stage process, where an initial alcoholic fermentation carried out by a yeast is followed by a bacterial acetification stage. The yeasts involved are traditionally those used for the production of alcoholic beverages peculiar to the country concerned. This reflects the origin of vinegar, as a means of utilizing excess or spoilt alcoholic beverages. The modern vinegar process includes the addition of *Saccharomyces diastaticus* to the fermentation to maximize carbohydrate assimilation and therefore removal. Species of *Acetobacter* are used for the second stage of the process wherein ethanol is converted to acetic acid.

The *Acetobacter* fermentation is widely carried out in two types of vessels. The more traditional trickling generator (Figure 2.6) is now being replaced by

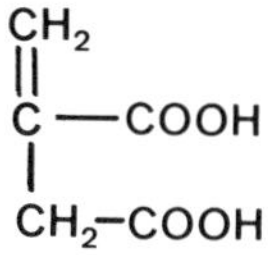

$$\begin{array}{c} CH_2 \\ \parallel \\ C-COOH \\ | \\ CH_2-COOH \end{array}$$

Figure 2.6 Itaconic acid.

the Frings acetator (Figure 2.7) (Nickol, 1979). More recently, a tower fermenter system has become available (see Chapter 4). It consists of a tubular fermenter with a 10:1 aspect (i.e. height:diameter) ratio. Liquor is allowed to trickle through the tower and is aerated by a countercurrent of air which is forced through plastic perforated plates covering the cross-section of the vessel.

2.4.4 *Polysaccharides*

A number of industrial development programmes are investigating the feasibility of replacing polysaccharides from plant and seaweed sources with microbial compounds. Development of novel, microbial polysaccharide-dependent processes is also under way. Polysaccharides have many industrial uses, arising from their capacity to alter the rheological properties of aqueous solutions. The resulting solutions may be Newtonian, pseudoplastic or thixotropic in nature (see Chapter 4). Large-scale industrial extraction of starch alginates, and to a lesser extent xanthan, provides raw materials for the detergent, textile, adhesive, paper, paint, food and pharmaceutical industries. The only microbial polysaccharide that has reached and maintained large-scale commercial production is xanthan gum, the polysaccharide produced by *Xanthamonas campestris*.

A potential market for microbial polysaccharides exists in tertiary oil recovery. Aqueous polymer solutions reduce the flow capacity of water pumped into a well, thereby increasing the efficiency of contact with and

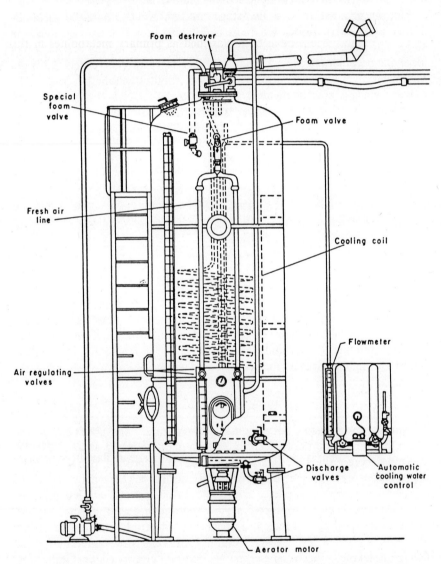

Figure 2.7 Frings acetator for vinegar production. From Nickol (1979), with permission.

displacement of oil. *X. campestris* polymer has been shown to produce higher viscosity and be less sensitive to saline waters than other polymers.

X. campestris is grown in highly aerated culture vessels of conventional design. A pH control system is necessary for this acid-producing fermentation. Problems encountered during the process are susceptibility to contamination, due to the bacterium's low substrate affinity, and the difficulty of maintaining high oxygen transfer rates in a culture fluid whose viscosity has been increased by excreted polysaccharide.

The polysaccharides have been classified as primary metabolites in this chapter. No unifying principles may be applied to production kinetics for polysaccharides, however, as they appear to range from those of primary metabolites to those of secondary metabolites, depending upon the polymer in question.

2.4.5 *Other primary metabolites*

The hematopoietic activity of vitamin B_{12} is utilized in the treatment of pernicious anaemia. Requirements for this function alone, however, cannot account for the demand for this compound. A proportion of the B_{12} produced throughout the world is used in the synthesis of other cobalamins, and pure preparations of the compound are sold as dietary additives, particularly in Japan.

Numerous microorganisms produce significant quantities of vitamin B_{12}. These include propionibacteria and *Pseudomonas* and *Streptomyces* species, all of which have been considered for industrial production. Resistance to cobalt has been used for selection of higher-producing mutants.

Propionibacterium denitrificans synthesizes etiocobalamin, a B_{12} intermediate, under anaerobic conditions, which it converts to deoxyadenosylcobalamin, a B_{12} precursor, in aerobic cultures. The production procedure is therefore a two-stage process with an appropriate aeration shift separating the first and second phases. In some processes a diphasic batch culture is employed, in others, a two-stage continuous culture.

Significant quantities (3000 tonnes, p.a.) of inosine monophosphate (5'-GMP) are required, mainly by the Japanese food industry, as flavouring agents. Production is carried out both by hydrolysis of yeast RNA and directly by fermentation. The two compounds have a flavour-synergistic effect with sodium glutamate in complex food seasonings in which 5'-IMP and 5'-GMP are added at levels of 10% and 2% respectively.

Fermentation methods for the production of 5'-IMP have been described which involve cultures of an adenine auxotroph of *Bacillus ammoniagenes* (Nakao, 1979). The fermentation profile resembles that of a secondary metabolite. Maximum 5'-IMP accumulation takes place when the culture is no longer growing at its maximum rate. Hypoxanthine accretion kinetics are characteristic of primary metabolite formation, and it has been suggested that this compound acts as a substrate for 5′-IMP-synthesizing enzymes which are produced when the culture substrate concentration drops below a critical level.

2.4.6 Single cell protein (SCP)

Single cell protein is crude or processed microbial biomass intended for human or animal consumption. Unlike extracellular metabolite production, which is often indirectly linked with cell growth, the formation rate of single cell protein is synonymous with biomass accretion rate. The kinetic models for microbial growth may therefore be utilized to describe and predict the behaviour of a single cell protein fermentation. The principal objective in designing such a process would appear to be operating at close to the maximum specific growth rate (eqn 2.3). The cell composition of many microorganisms, however, is a function of growth rate. Of particular relevance here are the intracellular protein and RNA concentrations. The protein concentration and amino acid composition effectively determine the nutritional value of the final product. However, a high nucleic acid intake in humans results in a high concentration of blood uric acid which, instead of being excreted, accumulates as crystalline deposits in the joints and soft

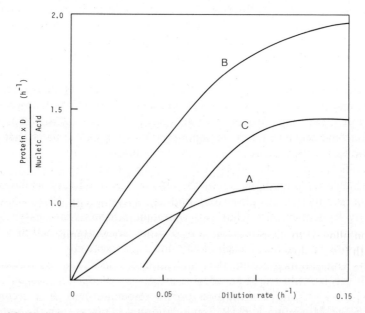

Figure 2.8 Variation of ribosomal efficiency with growth rate in (A) *Gibberella fujikuroi*, (B) *Saccharomyces cerevisiae*, (C) *Aspergillus nidulans*.

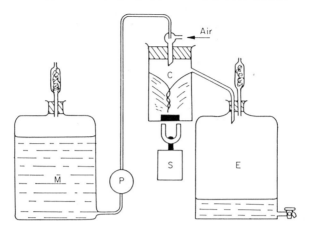

Figure 2.9 Essential features of a chemostat. Sterile medium (M), contained in a reservoir, is added to the culture (C) at a pre-determined rate by means of a metering pump (P). The culture is agitated with a magnetically coupled stirrer (S) and aerated. The culture volume is maintained constant by a constant level overflow device and effluent culture (E) is collected in a receiver bottle. From Tempest (1970), with permission.

tissues, leading to gout-like conditions. A growth rate which gives the highest protein to RNA ratio (Figure 2.8) may therefore be chosen for production.

Interest in SCP in the West has fluctuated with various socioeconomic considerations, but remains significant in Eastern bloc countries. There is some interest in developing SCP as an end product for cellulosic waste feed stocks.

2.4.6.1 *Continuous culture technique* (see 4.1.3). From the foregoing discussion, it will be apparent that in an ideal single cell protein fermentation process, specific growth rate could be controlled at an optimal value for a prolonged period. This may be achieved in continuous culture, a growth system which has become the method of choice for single cell protein production. Continuous culture is usually carried out in a device known as the chemostat (e.g. Tempest, 1970). The medium reservoir M (Figure 2.9) and the culture vessel C are filled with a growth medium with a single growth-limiting substrate (gls). A batch culture is then operated and the gls concentration allowed to drop to a value sufficiently low to significantly limit the growth rate. A slow feed of medium from the reservoir M is pumped into the culture, which can only then grow at a rate dictated by the rate of gls feed. Spent culture flows out of the vessel so that the volume is maintained at a constant level. The rate of input of medium, which is also equal to the rate of outflow of culture broth. The biomass concentration in the fermenter therefore remains constant: if the flow rate given by the pump is F ($l\,h^{-1}$) and the volume of the vessel is V (l) then the flow rate per unit volume will be F/V. This ratio is known as the dilution rate, D (h^{-1}).

We can deduce the relationship between D and the specific growth rate, μ (which also has units h^{-1}), by drawing up a mass balance:

Rate of change of biomass in a continuous culture	=	Rate of growth in the fermenter	minus	Rate of removal from the fermenter
$\dfrac{dx}{dt}$	=	μx	−	Dx

Since the biomass concentration in a chemostat is constant, when the steady state condition has been reached then $dx/dt = 0$ and $\mu = D$.

If biomass concentration is measured at a number of growth rates (Figure 2.10) it maintains a constant value (except at high and low dilution rates). As the value D approaches that of μ_{max}, the culture begins to wash out of the vessel. At extremely low growth rates, insufficient supply of gls is available for a significant biomass accretion rate as well as the 'maintenance' function (turnover of cell materials, preservation of osmotic gradients and cell motility), so the biomass concentration drops.

If the productivity Dx (g biomass l^{-1} h^{-1}) is plotted against dilution rate, an optimum occurs (Figure 2.11). Single cell protein plants are often operated at a compromise growth rate to achieve optimal productivity and a high protein to RNA ratio.

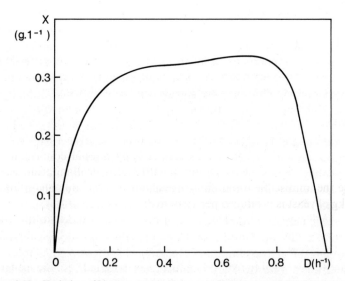

Figure 2.10 Variation of biomass $\bar{x}$ with dilution rate D in a glucose-limited chemostat.

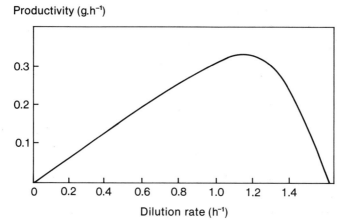

Figure 2.11 Variation of productivity $D\bar{x}$ with dilution rate in a glucose-limited chemostat.

2.4.6.2 *Development of processes for single cell protein production.* The most widely studied substrates are *n*-paraffins, which were chosen for most of the early processes. Work in this area was initiated during the development of a microbial process for dewaxing crude petroleum. Microorganisms capable of assimilating *n*-paraffins as carbon sources were employed. The high yield of biomass, together with predictions of an imminent protein shortage, caused a shift of emphasis in objectives whereby the production of microbial protein became the goal. The work was pioneered by BP, who constructed a 16 000 tons p.a. plant for a gas–oil substrate and a 4000 tons p.a. plant based on purified *n*-paraffins. The latter process was adopted after comparative pilot trials.

2.4.6.3 *BP type process.* A yeast, *Candida lipolytica*, was chosen for biomass production in this type of process. Although bacteria, in general, have higher efficiencies of carbon conversion, faster growth rates and higher protein contents, yeasts have the advantage of lower nucleic acid content, ease of separation and traditional acceptance as foodstuffs. *Candida lipolytica* was grown under carbon-limited continuous culture on a pure *n*-paraffin fraction in the C_{10}–C_{23} range. The fermenters are conventional continuous stirred tank reactors (CSTRs—see Chapter 4) with mechanical agitation serving to disperse the immiscible *n*-paraffin throughout the aqueous phase. At steady state, 1 kg of yeast is produced per kg of hydrocarbon feedstock.

2.4.6.4 *Shell type process.* This type of process is included to illustrate the use of a microbial community rather than a pure culture for an industrial process.

A continuous enrichment procedure was adopted for the isolation of methane-utilizing microorganisms. A soil-based inoculum was periodically

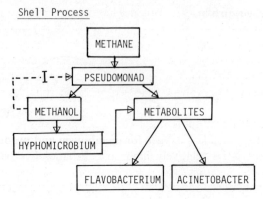

Figure 2.12 The Shell type of process for single cell protein production; microbial community for methane assimilation.

introduced into a methane-limited chemostat. It was found that a community consisting of a mixture of microbial species developed each time the experiment was performed. Selection of a monoculture using this procedure was never observed.

The most productive community (Figure 2.12) contained four recognizable species, only one of which was able to utilize methane for energy generation and biomass production. *Hyphomicrobium* benefits the methane-utilizing pseudomonad by removing its product, methanol, from the culture. As methanol is inhibitory to the pseudomonad, higher yields were obtained from this association than from the appropriate monoculture. Removal of *Flavobacterium* or *Acinetobacter* from the culture led to an unstable relationship. It is therefore thought that they act in removal of other self-inhibiting metabolites from the pseudomonad and *Hyphomicrobium*.

This community is nutritionally self-sufficient and it is therefore virtually impossible for a contaminant to grow in the culture (no further nutrients would be 'available'). It can also grow faster and provide a higher yield than any methane-utilizing monocultures tested. The community does not appear to produce foam in culture, presumably due to the lack of extracellular proteinaceous metabolites. Such desirable characteristics are likely to apply to other forms of mixed culture fermentation developed in future processes. Disadvantages of commercial-scale operation of this process included:

(i) the unpredictable price of soyabean and maize (competitors for the sale of single cell protein);
(ii) the potential in many areas of the world for further development of existing protein resources;
(iii) the difficulty of applying such a sophisticated process in less developed regions.

2.4.6.5 *Ethanol-based processes.* Ethanol has a number of attractive features as a fermentation substrate. It may be obtained as a very pure material, it is acceptable as a raw material for foodstuffs and it can be readily derived synthetically from ethylene. It is utilized by a much wider range of microorganisms than the hydrocarbon substrates so far considered; this may be considered to be a disadvantage, however, as it allows greater potential for contamination.

Unlike *n*-paraffins, ethanol is water-soluble, and, being in a higher oxidation state, requires much less oxygen for metabolism and produces less heat. The disadvantage is its relatively high cost. Ethanol has been used in a yeast-based process by the Amoco Company for the production of food products for human consumption.

2.4.6.6 *Pekilo process.* Waste materials, containing cellulose and other carbohydrates, have become more attractive as single cell protein fermentation substrates than petroleum-based feedstocks. Projected costs and increasing shortages of crude oil, problems in Italy and Japan, and growing interest in constructive utilization of waste materials have been contributory factors.

The Pekilo process, operating in Finland, utilizes sulphate waste liquor, a by-product of the wood pulp and paper industry. The organism is *Paecilomyces variotii*, a filamentous mould, and the product is sold as pig, chicken and calf foodstuff.

2.4.6.7 *Tate and Lyle type process.* Small-scale 'village-level technology' has been devised for underdeveloped countries. The small plastic tower fermenters are not operated aseptically. The substrates used are locally-available agricultural waste (carob-pods, papaya and olive, palm and potato waste waters) and the organisms are filamentous moulds (*Fusarium* and *Aspergillus niger*).

2.4.6.8 *Symba process.* This process, whereby the substrate is treated sequentially with cultures of different microorganisms, was developed jointly by the Swedish Sugar Company and Chemap (a fermenter manufacturer) of Switzerland. Starch in potato processing waters is first hydrolysed to low molecular weight sugar by *Endomycopsis fibuliger*. The culture is then inoculated with *Candida utilis*, which makes up the bulk of the biomass formed in this process.

2.4.6.9 *Rank Hovis McDougall type process.* This type of process was originally developed to utilize carbohydrate-containing waste from the flour milling industry. By addition of various growth factors to the medium, a high growth rate has been obtained for the organism, *Fusarium graminearum*, a filamentous mould. The product is sold in the UK as a constituent of pies and flans.

2.4.6.10 *ICI process.* For engineering aspects, see 4.2.2, 4.2.5 and 4.7.4. This process (Smith, 1980) was developed by ICI to exploit the availability of

Table 2.5 Advantages of methanol as a substrate for single-cell protein (SCP).

(i) It is completely miscible with water.
(ii) It can be produced from a very wide range of hydrocarbon feedstocks ranging from coal to naphtha, and also from methane in natural gas.
(iii) It can be produced in virtually unlimited quantities in any area of the world having any form of fossil fuel supplied and is not limited by the output of a refinery in the same way as normal alkanes.
(iv) Its method of production, which involves catalytic degradation of hydrocarbon feedstocks first to CO and H_2, precludes the presence of potentially carcinogenic polycyclic hydrocarbons.
(v) The growth microorganisms on methanol requires less oxygen than growth on normal alkanes—an important factor in the overall process economics.
(vi) If a methanol plant is built as an integral part of the SCP plant, considerable quantities of heat liberated by the methanol synthesis can be used to provide the sterilization heat required for the fermentation process.

---

methane from the North Sea gas fields. This can be converted to methanol by a chemical process, and bacteria are grown on the methanol to yield SCP suitable for animal feed. Methanol has a number of advantages (Table 2.5) as a fermentation substrate and the organism, *Methylophilus methylotrophus* has a very low K_s value (eqn 2.3) for methanol, so even at low concentrations this organism can use it efficiently as a substrate.

The production fermenter is very large, with a working volume of approximately 1.5×10^6 litres (see Chapter 4). The power requirements for agitating and aerating a vessel of this volume by stirring would be prohibitively high. An alternative system, the pressure cycle fermenter, has therefore been developed. Air and gaseous ammonia (the nitrogen source for the fermentation) are introduced at the base of the fermenter (Figure 2.13). The hydrostatic pressure generated by the weight of the column of liquid above causes circulation in the fermenter, and also increases gas solubility in this region. As the culture moves up the fermenter, the hydrostatic pressure decreases and the solubility of dissolved gases drops accordingly. To facilitate downstream treatment, however, the vessel is pressurized and the dissolved respiratory CO_2 is released from the harvest broth at a later stage in processing. The circulation cycle in the fermenter is apparently adjusted to match the cell division rate so that the region of high oxygen solubility coincides with maximum oxygen demand.

The rate of methanol addition is determined by the mass balance observed at the operating dilution rate.

$$1.72 CH_3OH + 0.23 NH_3 + 1.15 O_2 \rightarrow 1.0\,[\text{cells}] + 0.72 CO_2 + 2.94 H_2O.$$

The rate of oxygen consumption and carbon dioxide evolution are determined by gas analyses and the feed rate of methanol is regulated to maintain a respiratory quotient of $0.72 CO_2 / 1.15 O_2 = 0.48$.

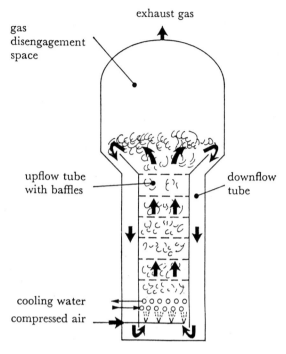

Figure 2.13 A pressure cycle fermenter. From Smith (1980), with permission.

The pressure cycle design of this fermenter achieves high rates of oxygen transfer and mixing without the problems of maintaining asepsis which arise when moving parts have to be sealed. The fermenter is operated at a dilution rate of $0.2\,h^{-1}$ with a steady state biomass concentration around $30\,g\,l^{-1}$.

The wild-type *Methylophilus methylotrophus* operates a two-stage pathway (Figure 2.14) for ammonia assimilation. This involves the enzymes glutamine synthetase (GS) and glutamate synthase (GOGAT) in a sequence which consumes one molecule of ATP and one of NAD(P)H per mole of ammonia fixed. The GS–GOGAT system has a high affinity for ammonia. A lower-affinity one-stage assimilation mechanism involving the enzyme glutamate dehydrogenase (GDH) exists in many microorganisms. GDH does not require ATP (Figure 2.14) and should, therefore, be more economical in terms of energy substrate utilization. The ammonia assimilation system is therefore likely to be a potential source of methanol wastage, and the production strain was subjected to a programme of genetic improvement. The GDH gene of *Escherichia coli* was cloned (Chapter 3) into broad host-range plasmids which complemented GOGAT-deficient mutants of *Methylophilus methylotrophus*. The recombinant organism was, indeed, found to have acquired a more

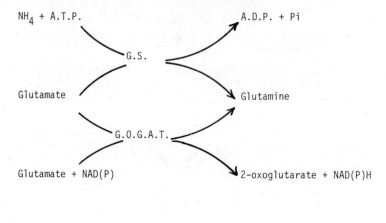

Figure 2.14 The GDH and GS-GOGAT systems for ammonia assimilation in bacteria.

energy-efficient ammonia assimilation mechanism during pilot studies, which produced 7% more biomass compared with the original strain, from the same amount of methanol. Thus, innovations in biochemical and genetic engineering and microbiology have been combined into an operating, commercially viable process. The resulting SCP was called Pruteen. Future developments are likely to exploit this fermentation technology in the development of novel processes for waste disposal (deep shaft fermenter) and for polyhydroxybutyrate production from microorganisms to make plastics.

2.4.7 Future plans in the UK

A consortium of companies and the Department of Trade and Industry are seeking to combine the advantages of the RHM mycoprotein product with ICI's production technology. The intention is, therefore, to produce *Fusarium graminearium* for human consumption in the 1.5 million litre air-lift fermenter.

2.5 Secondary metabolites in biotechnology

2.5.1 Penicillin

Alexander Fleming's discovery of penicillin in 1929 is part of microbiological folklore. A contaminant mould colony on a petri dish culture caused lysis of colonies of *Staphylococcus* growing on the agar. The implication of this observation was that a water-soluble antimicrobial compound has been excreted by the mould colony. This observation initiated the development of many other antibiotics. Penicillin remains, however, the most active and least

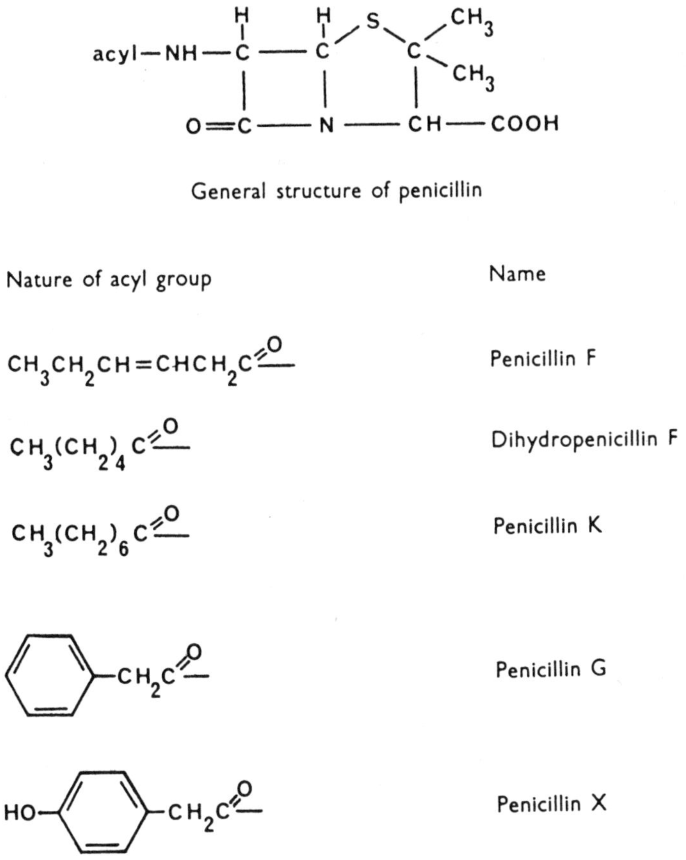

```
              H       H   S     CH₃
  acyl—NH—C ——— C       C
              |       |     \  CH₃
              |       |      |
           O=C ——— N ——— CH——COOH
```

General structure of penicillin

Nature of acyl group	Name
$CH_3CH_2CH=CHCH_2C{\overset{O}{\underset{}{\diagup}}}-$	Penicillin F
$CH_3(CH_2)_4C{\overset{O}{\underset{}{\diagup}}}-$	Dihydropenicillin F
$CH_3(CH_2)_6C{\overset{O}{\underset{}{\diagup}}}-$	Penicillin K
(phenyl)–$CH_2C{\overset{O}{\underset{}{\diagup}}}-$	Penicillin G
HO–(phenyl)–$CH_2C{\overset{O}{\underset{}{\diagup}}}-$	Penicillin X

Figure 2.15 The naturally-occurring penicillins.

toxic antibiotic in normal clinical use. Engineering aspects of penicillin production are discussed in 4.7.2, and genetic considerations in 3.2.5.

2.5.1.1 *The product.* The name penicillin denotes a number of substances produced by *Penicillium* and related fungi. Naturally-occurring penicillins have a common ring structure (Figure 2.15) to which different acyl side chains are attached.

Early penicillium fermentations produced mainly penicillin F, but addition of corn steep liquor (a by-product of the maize starch industry) to the growth medium not only improved the yield of total penicillins but had a qualitative effect, the major product becoming penicillin G. This was because corn steep liquor contains phenylacetic acid, a precursor of penicillin G. Addition of other compounds to the medium permitted the production of a large number

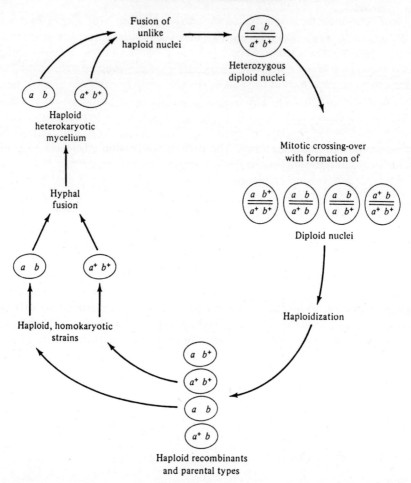

Figure 2.16 The parasexual cycle in *Penicillium chrysogenum*. From Wagner and Mitchell (1964), with permission.

of penicillins which are not produced naturally. A number of semi-synthetic penicillins are now produced which are manufactured by chemical conversion of penicillin G. Penicillin V is manufactured by direct fermentation.

2.5.1.2 *Penicillium chrysogenum.* The genus *Penicillium* belongs to the ascomycete class of the Eumycotina. The penicillia and aspergilli are the two most commonly occurring fungal genera. Penicillia form asexual spores known as conidia which make up a significant proportion of the aerial mycelium. Some penicillia also form perfect or sexual stages which result in the production of ascospores in a structure called the cleistothecium.

In *P. chrysogenum*, a parasexual cycle exists (Figure 2.16). Haploid vegetative hyphal fusion occurs, giving rise to mycelium containing a mixture of nuclei from the two strains. These then fuse, forming diploid nuclei which may persist for many mitotic divisions, and crossing-over may occur at a low frequency. Haploidization then takes place, leading to random re-assortment of the chromosome pairs. The process therefore differs from sexual cycles in that no meiosis occurs.

The parasexual cycle has been exploited along with mutagenesis to produce higher-yielding strains. The carbon conversion efficiences to penicillin of modern production strains are about 70%.

2.5.1.3 *Biosynthesis.* The primary metabolites valine, cysteine and lysene are precursors (Davis *et al.*, 1968) of penicillin. Valine and cysteine are direct precursors of the dihydrothiazine and β-lactam rings respectively. Since valine biosynthesis is under the feedback control of valine, the sensitive enzyme is acetohydroxyacid synthetase. It has been demonstrated that this enzyme is less sensitive to valine inhibition in high-producing mutants than in the lower-producing parent strain. Cysteine biosynthesis is rate-dependent upon the activity of the sulphate permease system, which is responsible for the assimilation of inorganic sulphur from the culture medium. Its activity has been found to be less sensitive to cysteine inhibition in enhanced production mutants. Lysine and penicillin are end-products of a branched biosynthetic pathway (Figure 2.17). This explains an observation that peni-

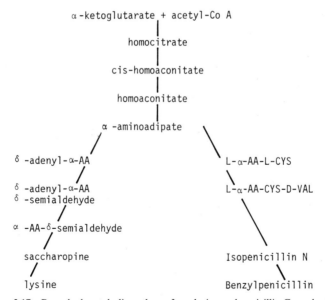

Figure 2.17 Branched metabolic pathway for L-lysine and penicillin G production.

cillin formation is subject to inhibition by lysine. Lysine causes feedback inhibition in the early, common part of the pathway.

L-Cysteine condenses with L-amino-adipic acid to form the dipeptide L-α-AA-L-cys. The addition of L-valine to this structure results in an optical inversion, to the D-form, of the valinyl moiety in the resulting tripeptide. Racemization of component amino acids to the D-form appears to be common in secondary metabolism. The Arnstein tripeptide, as the L-α-AA-L-cys-D-val is known, contains all the elements necessary for the synthesis of isopenicillin N. The mechanism of closure of the thiazolidine and β-lactam rings is still subject to investigation. Some available evidence suggests, however, that the nitrogen at position 4 is first oxidized, causing the spontaneous formation of the 4–5 bond, resulting in the β-lactam ring. The thiazolidine ring closure is probably initiated by oxidation of the sulphur.

In penicillin G formation, the acyl-transferase which effects the substitution of the phenylacetic acid has been isolated, and this has been demonstrated to be a rate-limiting factor in the whole process. Mutants with enhanced yields of penicillin G often show increased activity of this enzyme.

2.5.1.4 *Production.* In the phase of the process known as inoculum development, the overall objective is to provide an inoculum for a fermenter with a volume of some 2.5×10^5 l, from a freeze-dried ampoule of spores. The shape of each fermenter in the sequence may vary, however (Figure 2.18), and this may lead to non-uniformity of culture conditions. The important dimension to consider here is the aspect ratio. As the height varies, the hydrostatic pressure will also change. This will give rise to different values for oxygen solubility at the bottom of the fermenter at each stage. Each unit in the sequence is usually matched in terms of power input, from the stirrer, per unit

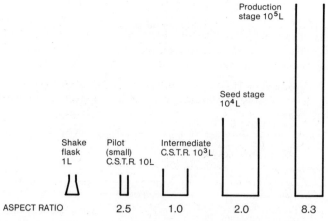

Figure 2.18 Variation in fermenter dimensions at different stages in inoculum development.

volume of culture. Because of the differences in geometry, however, differences in the sequence of physiological events in the cultures in the different stages may occur.

For this reason, the progress of cultures in each stage is monitored. Samples are taken from the fermenter at intervals for microscopic examination, and the progressive changes in culture pH, dissolved oxygen and biomass are checked. Biomass or cell growth is measured by the increase in broth viscosity, and this can be monitored by the changing power requirements of the stirrer motor to stir both to constant speed. If pH is being controlled, the rate of addition of pH control reagent to the culture may be used to estimate growth rate.

The inoculum is transferred to the next vessel once it is considered to be in mid-exponential phase of growth. It is transferred along pipes previously sterilized by the passage of steam. Sterility checks at each stage enable any source of contamination to be identified. The time taken to reach mid-exponential phase may vary between production runs. The plant manager estimates the time for transfer by the physiological parameters of dissolved oxygen, pH, and biomass, and arranges for the required manpower to be available. As this can be at any time of day or night, the personnel involved are invariably shift workers. Plant management may be simplified to some extent, however, in fully automated plants.

A further consideration is the determination of culture temperature. Growth, expressed as mycelial nitrogen, follows exponential kinetics over the first 24 h in this example. pH varies at will between 5.5 and 8.0, and the carbohydrate or sugar utilization rate follows a diauxic curve. Penicillin production rate becomes maximal between 30 and 50 h and degenerates thereafter.

The culture temperature would have been chosen by running this entire process at a number of different temperatures and selecting the temperature which gave the highest overall yield of penicillin. The effect of temperature may not be consistent for individual process variables, however (Table 2.6). Thus the optimum temperature required to achieve maximum cell density (and maximum production capability) is 30°C. Instead of running the process at a single compromise temperature, therefore, the fermentation is operated according to a temperature profile (Figure 2.19) whereby the

Table 2.6 Influence of culture temperature on growth and productivity in *Penicillium chrysogenum*.

Parameter	Optimum temperature
Biomass accretion	30°C
Respiration rate	22–29°C
Penicillin production rate	25°C

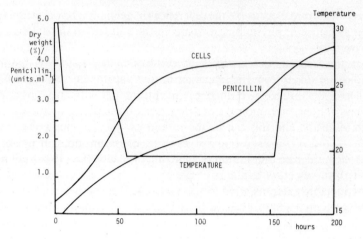

Figure 2.19 The use of a temperature control profile for penicillin production in *Penicillium chrysogenum* cultures.

optimum temperature requirements for each culture phase are maintained.

For the first five hours, the temperature is set for maximum growth rate. Between 15 and 45 h, penicillin biosynthesis starts and the temperature is altered to the optimum value for antibiotic production. At 45 h, the temperature-dependent rate of penicillin destruction in the culture becomes the rate-limiting influence on penicillin accretion, and the temperature is lowered accordingly. After 140 h, the penicillin destruction rate appears to be less important and the temperature is returned to that which favours antibiotic biosynthesis.

The use of such a profile depends upon reproducibility in successive production runs. This limitation may be overcome by applying computer control; the controlling algorithm would be capable of adjusting the length of the different temperature regimes on the basis of information received from the fermenter. The rate of change of power input would indicate the achievement of maximum cell density. Demand for pH control, oxygen and on-line automatic product determination could be used to determine the subsequent phases.

The same principles apply to pH profiles which have been developed for this fermentation.

The rate of carbohydrate uptake is maximal during the first 30 h of the fermentation, which corresponds to the phase of exponential mycelial growth. At the minimum value for carbohydrate assimilation (40 h), maximum penicillin biosynthesis occurs. This inverse relationship between sugar uptake and antibiotic production is consistent with catabolite repression of biosynthesis. As the uptake rate increases, presumably the intracellular carbohydrate concentration rises, and the rate of synthesis decreases.

A transitory phase of maximum product formation occurs (Atkinson, 1974). In order to maintain the penicillin production rate at an optimum level, it is necessary to control the carbohydrate uptake rate. In order to exploit this phenomenon, the fed batch culture has been developed. In its least sophisticated form, enough carbohydrate is added at the start of the fermentation to ensure rapid mycelial growth and then, at about 40 h, a sugar feed is introduced to the fermenter throughout the production phase. The feed is first introduced at a time which coincides with the exhaustion of the carbohydrate added at the start of the process. In some processes, a sugar feed of variable rate is added throughout the fermentation. This practice follows the observation that the rate of product formation decay in the culture depends upon the maximum specific growth rate attained during exponential phase. To minimize product formation rate decay, therefore, a high growth rate must be achieved (see Pirt and Righelato, 1967). Suitable growth rate control may be effected by a growth-limiting feed of carbon source during the exponential phase.

The optimum carbohydrate feed rate for antibiotic production phase may be accurately controlled by monitoring the dissolved oxygen uptake, since the two parameters have a linear relationship (Righelato et al., 1968).

During the phase of antibiotic synthesis, the pH value of the culture tends to rise as ammonia is liberated by deamination of nutrient amino acids. This occurs as a result of glucose exhaustion. Instead of adding acid to restore the pH to its predetermined value, it is possible to control culture pH by addition of glucose to the medium in an equimolar quantity to the ammonia liberated. This may be achieved by linking the glucose feed pump to a pH monitor and represents an alternative method of carbohydrate addition to that outlined above. This approach could also form the basis of a computer control algorithm.

Phenylacetic acid, the side-chain precursor to penicillin G, is added to the culture in the form of a slow feed, as it is toxic to *P. chrysogenum* in high concentrations. Trial and error experimentation has demonstrated that the optimal process regime is a feed at 1.5 times the glucose concentration in the carbohydrate feed.

2.5.1.5 *Harvest*. The decision to terminate the fermentation may be based upon one of four parameters:

(i) Rapid chemical antibiotic assays performed on culture samples
(ii) Reduction in the culture glucose demand
(iii) Reduction in culture dissolved-oxygen concentration
(iv) A characteristic change in the pH trend.

(i) is suitable for penicillin, as a rapid assay is available. Productivity can, however, follow a diauxic (two-peak) pattern. A reduction in penicillin production rate does not, therefore, in itself indicate the best time to stop the fermentation. Parameters (ii) and (iii) indicate that the cell density has exceeded the oxygen transfer capacity of the culture vessel so that the fermentation has become oxygen-limited, and (iv) indicates that cell lysis is beginning as a result of oxygen limitation. The decision to harvest may be taken on the basis of a combination of (i) and one or more of the other parameters.

The production principles for penicillin G have been applied to the biosynthesis of a number of other secondary metabolites as well as many primary metabolites, but blind acceptance of the universal applicability of such principles has resulted in many suboptimal fermentation development programmes. A hallmark of the more successful new fermentation processes is the development of fundamental new approaches, individually tailored to the product in question.

The penicillin fermentation is, nevertheless, the most highly developed process for secondary metabolite synthesis and has been discussed in some detail here to illustrate approaches which may be used in the optimization of antibiotic production.

2.5.2 Other secondary metabolites

2.5.2.1 *Cephalosporins.* A class of β-lactam antibiotics, known as the cephalosporins (Figure 2.20) is obtained from the filamentous mould *Cephalosporium acremonium* and a number of streptomycetes. Cephalosporins are often co-produced with penicillin N and other antibiotics. Like the penicillins, they are derived from the Arnstein tripeptide.

Methionine acts as an inducer for cephalosporin C production, exerting its effect when added during the growth phase, before antibiotic synthesis commences. In *C. acremonium*, methionine stimulates cysteine desulphhydrase

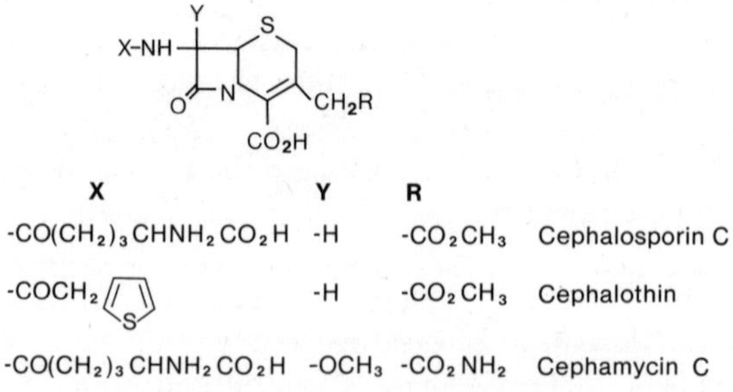

X	Y	R	
$-CO(CH_2)_3 CHNH_2 CO_2H$	-H	$-CO_2CH_3$	Cephalosporin C
$-COCH_2$ (thiophene)	-H	$-CO_2CH_3$	Cephalothin
$-CO(CH_2)_3 CHNH_2 CO_2H$	$-OCH_3$	$-CO_2NH_2$	Cephamycin C

Figure 2.20 The cephalosporins.

(L-cystathione→L-cysteine) and L-serine sulphhydrase (L-homoserine + L-serine→cystathione) resulting in the accumulation of L-cysteine, a precursor of cephalosporin C. The many synthetic derivatives of cephalosporin C include the cephamycins, which are resistant to β-lactamase enzyme attack.

2.5.2.2 *Aminoglycosides.* Aminoglycosides are bases containing amino sugars, the best known of which is streptomycin. It is produced by the actinomycete, *Streptomyces griseus*, which was the first commercially important species to emerge from a soil screen for novel antibiotic-producing microorganisms. Streptomycin production follows the normal pattern for secondary metabolite synthesis, taking place immediately after exponential phase. The process requires high levels of oxygen transfer and there are different optimum phosphate uptake rates for growth and antibiotic synthesis.

All three components of the structure are derived from glucose. Eight enzymes effect a bioconversion of scylloinosamine to streptidine, all of which are antibiotic-committed enzymes. The remainder are enzymes of primary metabolism. Arginine, glutamine and alanine also act as precursors, providing amino groups at various stages in the pathway (Claridge, 1979). Streptose is formed entirely from glucose by a rearrangement. Little is known about the individual steps in the conversion. Similarly, the N-methyl-L-glucosamine component arises from D-glucose and L-methionine, but the actual pathway is unknown.

Mannosidostreptomycin, a derivative of streptomycin, is formed concurrently with streptomycin until the enzyme mannosidostreptomycinase is formed later in the fermentation. The enzyme rapidly cleaves the compound to form streptomycin. It is induced by mannan, a component of yeast extract, which occurs in the production medium. The pH and temperature of the culture are adjusted to those optimal for mannosidostreptomycinase activity during the later part of the process.

2.5.2.3 *Macrolide antibiotics.* Macrolides are compounds with a macrocyclic lactone ring (Figure 2.21). Antibiotic forms have many carbons in the ring. The only macrolide in current therapeutic use is erythromycin, although

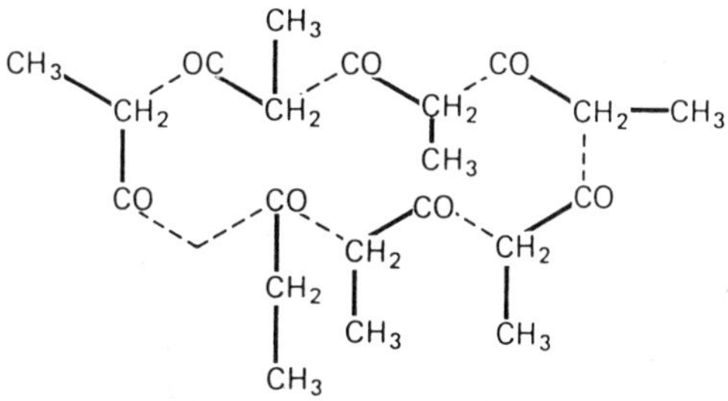

Figure 2.21 Erythromycin.

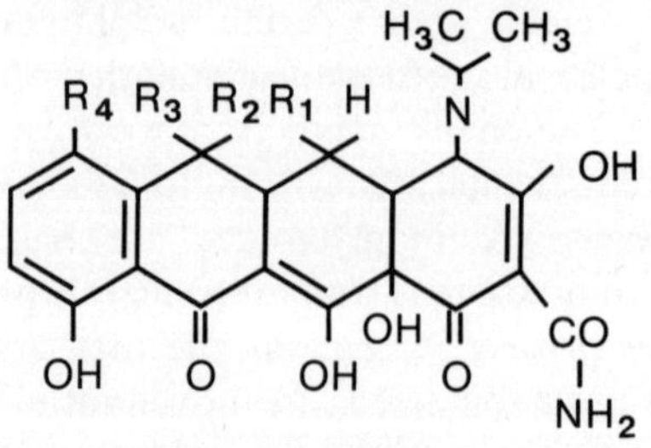

Figure 2.22 Tetracycline.

many more exist, and research programmes for the development for new structures are in progress.

Erythromycin is produced by *Streptomyces erythreus* from priopionyl CoA and methylmalonyl CoA precursors in a metabolic pathway which is analogous to that of fatty acid biosynthesis.

2.5.2.4 *Tetracyclines.* Tetracyclines were the first antibiotics to be discovered which could be administered orally. Aureomycin (chlortetracycline) (Figure 2.22) was discovered during a soil screening programme. It is produced by *Streptomyces aureofaciens*. Chlortetracycline is produced when a threshold concentration of chloride ions in the medium is exceeded, otherwise tetracycline is produced.

A lag phase of 12 h precedes exponential growth, and maximum nucleic acid synthesis takes place during this phase. The nucleic acid synthesis rate is enhanced by increased phosphate concentration in the medium. High nucleic acid biosynthesis rates lead to low eventual tetracycline production rates, however, so the phosphate concentration is kept low. Protein synthesis is also maximal during this period. TCA (Krebs) cycle enzymes, fatty acid synthesis and other enzymes of intermediary metabolism are maximal during exponential phase. The Embden–Meyerhof–Parnas (glycolysis) pathway is operative for glucose catabolism. Anaplerotic CO_2 fixation (to replenish the Krebs cycle), and a switch to glucose catabolism by oxidation in the hexose monophosphate shunt, characterize the biosynthesis process (probably both are results of running out of substrate). The carbon skeleton is derived from acetate units and methionine.

2.5.2.5 *Other β-lactam antibiotics.* Subsequent to the development of the cephalosporins a number of β-lactam antibiotics with novel structures were discovered. A new monocyclic antibiotic (nocardicin) has been discovered in *Nocardia uniformis*, an actinomycete. It is important as it is the first β-lactam to be discovered in this genus. It has a completely novel structure, and biosynthetic data indicate that the β-lactam ring is synthesized without the Arnstein tripeptide intermediate.

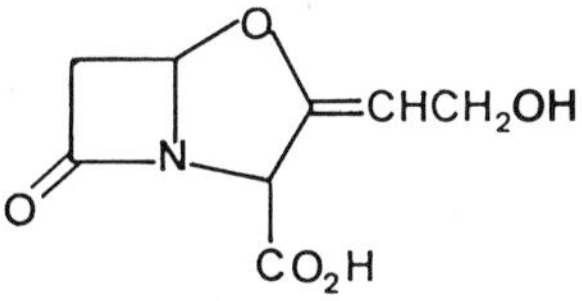

Figure 2.23 Clavulanic acid.

Clavulanic acid is a member of a group of natural substances that have been detected by their ability to inhibit the destruction of penicillin antibiotics by bacterial penicillinase. It is produced by *Streptomyces clavuligerus*, which also produces penicillin N and a number of cephalosporins. *S. clavuligerus* had been studied in connection with cephalosporin synthesis but clavulanic acid was not detected by normal antibacterial assays. It has a broad antibacterial spectrum but its level of activity alone is relatively low.

It has a unique fused oxazolidine ring structure (Figure 2.23). The clinical potential of clavulanic acid lies in the possibility of combining it with penicillins and cephalosporins to provide a preparation which will kill resistant bacteria.

The screening procedure which detected clavulanic acid has also detected a series of compounds known as olivanic acids. These antibiotics have been shown to be produced by a number of streptomycetes. Like clavulanic acid, their antibacterial activity is allied with their ability to protect other β-lactam antibiotics from β-lactamase activity.

Thienamycin resembles the olivanic acids in structure (Figure 2.24). It has been isolated from *Streptomyces cattleya*. Unlike clavulanic and olivanic acids, thienamycin is an antibiotic in its own right. It is 100 times more active than cabenicillin (a clinical penicillin) and has a much broader spectrum of activity, including pseudomonads and anaerobic pathogens. Thienamycin itself is unstable, but the development of stable derivatives should lead to the availability of an important new chemotherapeutic agent.

2.5.2.6 *The products of plant cell culture.* The development of the technology of plant cell tissue culture has resulted in the possibility of producing the metabolites of higher organisms, using microbial product formation techniques. Some of the principles for microbial product metabolite formation

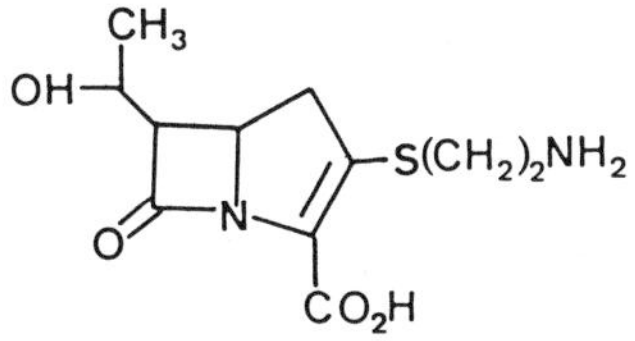

Figure 2.24 Thienamycin.

Table 2.7 Secondary metabolites detected in plant tissue cultures. (Adapted from Butcher, 1977.)

Compound	Type of culture
Cinnamic acids and derivatives	Mostly callus
Benzoic acids	Callus
Coumarins	Mostly suspension
Flavones and flavonols	Mostly suspension
Chalcones and deoxyflavones	Mostly callus
Coumestones and coumarinochromans	Callus and suspension
Anthocyanins	Callus
Tannins and tannin precursors	Callus and suspension
Anthraquinones	Mostly callus
Naphthoquinones	Callus
Sesquiterpenes	Mostly callus
Sterols and triterpenes	Callus
Steroidal alkaloids	Callus
Carotenoids	Callus
Unusual fatty acids and related compounds	Callus
Glucosinolates	Callus and suspension

may be applied to plant cells for the production of therapeutic and diagnostic agents, drug precursors and flavouring agents. These include allergens, disogenin, L-DOPA, ginenosides and glycyrrhizin. The known plant tissue culture secondary metabolites are listed in Table 2.7. Fermentation times are higher than those required by microorganisms by at least an order of magnitude. To avoid cell damage, gentle stirring is needed, and novel fermenter systems of the air-lift type are under development.

Bibliography and references

Atkinson, B. (1974) *Biochemical Reactions*. Pion, London.
Ayres, J. C., Mundt, J. O. and Sandine, W. E. (1980) *Microbiology of Foods*. Freeman, San Francisco.
Butcher, D. N. (1977) In *Applied and Fundamental Aspects of Plant Cell Tissue and Organ Culture*, eds. J. Reinert and Y. P. S. Bajij, Springer, Heidelberg.
Claridge, C. A. (1979) In *Secondary Products of Metabolism*, ed. A. H. Rose, Academic Press, London.
Davis, B. D., Dulbecco, R., Eisen, H. N., Ginsberg, M. A. and Wood, W. B. (1968) In *Principles of Microbiology and Immunology*, Harper International, New York.
Hirose, Y. and Okada, H. (1979) In *Microbial Technology*, vol. 1, eds. H. J. Peppler and D. Perlman, Academic Press, New York.
Nakao, Y. (1979) In *Microbial Technology*, vol. 1, eds. H. J. Peppler and D. Perlman, Academic Press, New York.
Nickol, G. B. (1979) In *Microbial Technology*, vol. 2, eds. H. J. Peppler and D. Perlman, Academic Press, New York.
Pan, C. H., Hepler, L. and Elander, R. P. (1972) *Devel. ind. Microbiol.* **13**, 103.
Pirt, S. J. and Righelato, R. C. (1967) *Appl. Microbiol.* **15**, 1284.
Righelato, R. C., Trinci, A. P. J. and Peat, A. (1968) *J. gen. Microbiol.* **50**, 399.
Smith, S. R. L. (1980) *Phil. Trans. R. Soc. London*, Ser. B, **290**, 341.

Tempest, D. W. (1970) In *Methods in Microbiology*, eds. J. R. Norris and D. W. Ribbons, Academic Press, New York.
Wagner, R. P. and Mitchell, H. K. (1964) In *Genetics and Metabolism*, John Wiley, New York.
Zahner, M. (1977) In *Antibiotics and Other Secondary Metabolites: Biosynthesis and Production*, eds. R. Hutter, T. Leisenger, J. Nuesh and W. Wehri, Academic Press, London, 1–17.

Supplementary reading

Industrial Applications of Microbiology, J. Riviere, M. O. Moss and J. E. Smith, Surrey University Press, Glasgow and London.

The Filamentous Fungi, vol. 2, J. E. Smith and D. R. Berry, Edward Arnold, London.

Annual Reports on Fermentation Processes, vols. 1 and 2, D. Perlman, Academic Press, London.

Biochemical Engineering, S. Aiba, A. E. Humphrey and N. Millis, Academic Press, London.

Principles of Microbe and Cell Cultivation, S. J. Pirt, Blackwell Scientific, Oxford.

Topics in Enzyme and Fermentation Biotechnology, ed. A. Wiseman, vol. 6, Chapter 2, Ellis Horwood, Chichester.

3 Applications of the principles of microbial genetics to biotechnology
JEREMY W. DALE

The main aim of the microbial geneticist in the field of biotechnology has been to alter the genetic composition of industrially important microorganisms so as to increase the overall efficiency of the process for which that organism is used. Although the primary requirement may be to increase the yield of a specific product, other parameters are also important, particularly improvements in growth characteristics and the elimination of undesirable by-products.

To this first aim, strain development, can be added two others: firstly, the establishment of strains which produce new or modified products, principally by the use of the more recently developed techniques of genetic engineering, and secondly, the application of genetic techniques has a crucial role to play in the understanding of the biosynthetic pathways leading to the production of commercially important metabolites, and of the control mechanisms that govern those pathways. Although such information could be very useful in strain development programmes, research of this nature has been relatively neglected by the applied microbial geneticist, and is regarded here as outside the scope of this chapter.

3.1 Control mechanisms in microorganisms (see also 2.3.3)

In order to appreciate the applications of genetic techniques to strain development, it is necessary to understand the mechanisms which may control product formation, and which therefore need to be exploited or eliminated.

The simplest case is that of production of an enzyme (or some other protein) which is not subject to induction or repression; that is, the level of production of that enzyme is not affected by the presence, or absence, of specific substrates or products in the medium. Such an enzyme is often described as constitutive, but it is wrong to assume that this means that it is always produced at its maximum potential level. The cell will have widely different requirements for different proteins, even where those needs are not affected by external conditions. Furthermore, the levels of so-called constitutive enzymes may be affected by non-specific changes in conditions; in particular, the levels of many enzymes are affected by changes in the rate of growth of the organism.

In addition to these constitutive enzymes, there are many other enzymes that are specifically regulated in response to external conditions. This applies particularly to those enzymes that are involved with specific degradative or synthetic pathways: enzymes involved in breakdown of lactose are not produced unless lactose is present, while those enzymes needed for the synthesis of histidine are only made if there is little histidine present in the medium. These are referred to as inducible and repressible systems: in the first case, lactose acts as the inducer, and in the second case expression is repressed by histidine.

The main parameters that affect the levels of enzyme production, which are therefore susceptible to mutational improvement or to deliberate manipulation by recombination *in vitro*, are as follows.

(1) *The number of copies of the gene concerned.* An increase in gene copy number would be expected to lead to an increase in the level of the product of that gene. Although this usually happens, the increase is often not proportional to the number of copies of the gene.

(2) *The efficiency of the promoter.* Every gene, or every collectively transcribed group of genes (forming an operon) is linked to a DNA region known as the promoter. This is the site which is recognized by the RNA polymerase prior to initiation of transcription. Promoters vary considerably in their efficiency, and there is therefore a considerable variation in the level of transcription of different genes. Since the interaction between the RNA polymerase and the promoter also depends on the specificity of the polymerase, a good promoter in one organism may not function effectively if that gene is transferred to another species. This is particularly important for *in-vitro* recombinant work, where genes may be transferred to totally unrelated organisms.

(3) *The presence of attenuators.* An attenuator is a transcription termination site within an operon which leads to a reduced level of transcription of genes beyond that site. An attenuator may also occur in the leader region of a gene, i.e. the region between the transcription start site and the structural gene itself. Attenuators are involved in the control of expression of a number of the operons involved in amino acid biosynthesis in prokaryotes; the absence of the amino acid concerned is thought to cause ribosomes to stall at a specific point on the mRNA. This in turn causes changes in the secondary structure of the mRNA which enable the RNA polymerase to proceed through the termination (attenuator) site. For further details, see Galloway and Platt (1986) or Platt (1986). Since this effect can be influenced by the availability of ribosomes, it can also provide a possible explanation for the enhanced

synthesis of many enzymes at high growth rates, when ribosomes are comparatively abundant.

(4) *Induction*. As explained above, this occurs particularly with degradative enzymes such as beta-galactosidase (in *E. coli*), which is not produced unless lactose or some other suitable inducer is present in the medium. This effect is due to a protein repressor which binds to the operator site (a region which overlaps with the promoter) of the *lac* operon to prevent transcription. The presence of the inducer results in an allosteric change in the conformation of the repressor so that it is unable to bind to the operator. Constitutive mutants, which produce the enzyme fully in the absence of the inducer, may arise from alteration in the repressor gene to abolish repressor production or function, or by alteration of the operator site so that the repressor molecule is unable to bind. This is an example of a negative control mechanism, as opposed to a positive control system where the regulator protein, in combination with an inducer, stimulates transcription.

(5) *Repression*. In the case of enzymes involved in biosynthetic pathways, the requirement is for enzyme production to occur only in the *absence* of the specific product of that pathway. This can be achieved by a modification of the above model so that the regulator protein combines with the operator site on the DNA (to switch off transcription) only in the presence of an effector molecule such as the final product of the pathway concerned. This occurs, for example, with the tryptophan operon of *E. coli*, although in this case (as for other biosynthetic pathways) attenuation is also known to have a role in regulation.

(6) *Catabolite repression*. This differs from the above phenomena in being less specific in its effects. Once again, the *lac* operon of *E. coli* provides the best-known example. Lactose is unable to function as an inducer of beta-galactosidase (and the other enzymes of the *lac* operon) in the presence of a readily-utilized carbon and energy source such as glucose. In the presence of glucose, ATP levels within the cell increase, and the level of cyclic AMP (cAMP) decreases. For transcription of the *lac* operon to proceed, it is necessary for cAMP to combine with a cAMP receptor protein and for this combination to interact with the operator/promoter region of the *lac* operon. The low cAMP levels that exist when glucose is available prevent this interaction from occurring. A number of other genes will be affected at the same time.

(7) *Translational controls*. In some cases, the level of a protein may be limited by regulation of translation rather than transcription. Much less is known about translation control mechanisms, and it would be expected that they would be less important under normal circumstances, since it is 'wasteful' for the cell to produce mRNA in excess of the level required for synthesis of that protein. However, under circumstances where transcriptional controls

have been abolished by mutation, or where foreign or synthetic genes have been inserted by cloning, translational effects may limit the amount of protein produced. This may happen, for example, via the nature of the ribosome binding site on the mRNA, the availability of ribosomes, or the match between the codons used and the availability of the corresponding transfer RNA species.

(8) *Post-translational modification.* The primary translation product is often subjected to a variety of alterations to yield the final mature protein. These modifications include simple alteration of specific amino-acid residues, glycosylation, and cleavage of the 'signal peptide' that is involved in transport of secreted proteins across the cell membrane. Although such modifications do not usually affect the total amount of that protein that is produced, they are nevertheless often important in establishing a biologically active product, and also contribute to the overall control of the metabolism of the cell by altering the activity of specific enzymes.

So, if we are interested in maximizing production of a specific protein, there are several control mechanisms that may be amenable to mutation or other manipulations. Extending the argument to consider the products of either primary or secondary metabolism introduces two further significant factors. Firstly, more than one enzyme will usually be involved, and secondly, a cell regulates its metabolic activities not only by altering the *amount* of enzymes produced, but also by modifying the *activity* of the enzymes involved. This may occur by activation, or (more commonly in the synthetic pathways with which we are concerned) inhibition by the end-product of the pathway (feedback inhibition).

Let us consider the hypothetical pathway (Figure 3.1) for the production of a secondary metabolite S. This has two steps in common with the primary metabolite P. In this case, the first enzyme of the joint pathway is susceptible to both feedback inhibition and repression by P and by S. In addition, the enzyme (4) which converts the branch point compound C into S is inhibited and repressed by S. The following possibilities then exist for maximizing the production of S.

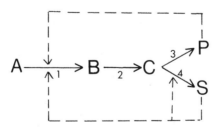

Figure 3.1 Hypothetical pathway for production of a secondary metabolite. ———→ feedback inhibition/repression; P, primary product; S, secondary product.

(i) Increasing the level of enzyme 4, by alteration of the control mechanisms described above. A mutation leading to an increase in the molecular activity of the enzyme itself (although a possibility) is much less likely.

(ii) Abolition, or reduction, of either the level of production or the molecular activity of enzyme 3. This would decrease the conversion of C to P, making more of the branch-point compound available for conversion into the desired secondary metabolite S. This has further consequences for the cell. If P is an essential substrate for other biosynthetic processes (if it is an amino acid, for example), the cell will now be auxotrophic, and the growth medium will need to be supplemented with that amino acid. However, since the first enzyme of the joint pathway is inhibited and repressed by P, addition of this to the medium may paradoxically abolish the production of S altogether. This may be overcome by a further mutation (iii).

(iii) Mutation of the gene for enzyme 1 to render the enzyme insensitive to feedback inhibition, or alteration of the control region for this gene so that it is no longer repressible. This would lead to an increased conversion of A to B, and consequently an increased production of C. This can also be achieved by an increase in the production of enzymes 1 and 2. However, since enzyme 4 which catalyses the production of S from C is inhibited and repressed by S, an increase in the rate of production of C will lead to an increase in P or to an accumulation of C, depending on the level of enzyme 3.

(iv) This can therefore be overcome in turn by mutation of the gene for enzyme 4 (or its control region) so as to maximize the conversion of C to S even in the presence of high levels of the secondary metabolite.

It can by now be appreciated that since this hypothetical pathway is a highly simplified one compared to that for the synthesis of, say, penicillin, the genetics of strain improvement is indeed a complicated subject.

3.2 Manipulations *in vivo*

The mainstay of genetic techniques *in vivo* is the process of selection. In bacterial genetics in particular, the term 'selection' is commonly used in a somewhat specialized sense, in that the population of bacteria is subjected to conditions under which only the required variants are able to survive and grow, i.e., selective conditions. Since the population may be very large (10^9 cells per ml can be readily obtained with many bacteria), variants that occur at very low frequencies (of the order of 10^{-9}) can be isolated if sufficiently powerful selective conditions are available. This would be the case for example with many types of antibiotic resistance, ability to use specific substrates, or resistance to a particular bacteriophage. If selective conditions are not available, the bacterial geneticist has to pick individual colonies from a non-selective medium and screen them for the required characteristic. This is a much less rigorous process, and even with rapid techniques such as replica plating or testing the colonies *in situ*, it would clearly be very laborious to isolate mutants that occur at such low frequencies. And yet, in most instances of commercially important strain improvement programmes, selective conditions in the above sense are not available, and it is therefore precisely this type of screening process that has to be employed. Rather confusingly, the term 'selection' is also used in this context in the more general sense of identifying and *choosing* variants with improved characteristics.

3.2.1 *Mutation*

A full discussion of the nature of mutation is outside the scope of this chapter. Briefly, however, a mutation may result from a variety of changes in the structure of the DNA, such as:

(i) A change in a specific base-pair (base substitution), which may result in a change in amino acid sequence or in polypeptide chain termination
(ii) Addition or deletion of one or two bases (frameshift)
(iii) Deletion of longer stretches of DNA
(iv) Insertion of DNA within a gene, which destroys gene function
(v) Other rearrangements of DNA such as transposition and inversion, and, in eukaryotes, chromosome rearrangements.

An elementary knowledge of protein synthesis and enzyme function will show that in the vast majority of cases these changes will be deleterious to the function of that gene or its product, and that it is therefore comparatively easy to obtain mutants that are defective with regard to a specific enzyme. It is far more difficult to get variants which produce an enzyme with enhanced molecular activity or altered specificity. However, increased *production* of an enzyme (as opposed to increased *activity* of the enzyme molecule) can be obtained by mutational alteration of the control mechanisms that may restrict production of that enzyme in the wild-type organism.

Microbial populations are normally remarkably homogeneous. A population of 10^9 bacteria contains far less genetic variation than an equivalent population of human beings. However, spontaneous variation does occur, albeit at a low level, due to the occasional 'errors' inherent in the process of DNA replication that escape the proof reading, editing and repair mechanisms that are responsible for maintaining the fidelity of the system. Given sufficient resources and patience, and a rapid testing method, advances can be made simply by screening large numbers of colonies. The initial improvements in penicillin production (see below) were achieved in precisely this manner.

The whole process can be speeded up quite considerably by using *mutagenic agents* to increase the proportion of mutants in the population. There are two main categories of mutagens: irradiation and chemical mutagens. Ultraviolet irradiation, which is commonly used, is only indirectly mutagenic. The primary effect is to cause the formation of pyrimidine dimers, i.e. the creation of a covalent linkage between adjacent thymine or cytosine residues on the same DNA strand. These pyrimidine dimers are lethal unless they are repaired. A number of repair mechanisms have been identified in *E. coli*, most of which are 'error-free' mechanisms which restore the original DNA sequence with at least the same degree of accuracy as the normal methods of DNA replication. Mutations arise from the operation of an 'error-prone' repair pathway with which the chances of incorrect bases being inserted are considerably increased. Exposure to x-rays has also been used, in some cases with great effect, to increase the proportion of variants, although it is much less convenient than ultraviolet irradiation and produces a considerable amount of gross chromosomal damage.

Among chemical mutagens, nitroso compounds, in particular N-methyl-N'-nitro-N-nitrosoguanidine (MNNG) have been widely used. MNNG produces 0–6 alkylation of guanine residues, and its mutagenic action may be due to misreplication. It has a preferential effect at the DNA replication point and therefore tends to produce clusters of closely linked mutations. Although this is in general undesirable since it results in the accumulation of mutations additional to the required one, co-mutation can be used as an aid to the isolation of mutants which are not themselves readily identified, by selecting mutants of a closely linked gene. As well as having the disadvantages of producing multiple mutations, MNNG is hazardous to use. A safer, although less potent, mutagen that also acts as an alkylating agent is ethylmethane sulphonate (EMS).

As well as these established methods of mutagenesis, there are a number of other causes of variation in microorganisms, arising from larger-scale genetic rearrangements and acquisition of new genes. Extrachromosomal elements such as plasmids and specialized transducing phages can be used for the transfer of genes between different strains, and for increasing the gene copy number. Many bacterial plasmids also contain DNA sequences known as *transposons*. These are regions of the DNA which include one or more recognizable genes and are able to promote their own transposition to another plasmid in the same cell or to the bacterial chromosome. This not only causes a rearrangement of the genes carried by the transposon itself, but also destroys the function of the gene within which the transposon becomes inserted. For more information on transposons, see Bennett (1985). Inversion of a DNA segment which controls the activity of a gene or a group of genes may also give rise to variable characteristics. This has been shown to be responsible for the phase variation of *Salmonella* flagellar antigens, for example. One of the characteristics of this type of variation is that the variant strains, in the absence of any selective pressure, tend to be unstable, and that the frequency of variation tends to be similar in either direction. Furthermore, the inversion may be catalysed by the products of other genes within the cell (see Smith, 1985).

Another type of variation is seen where a gene is present in the cell in two or more versions, only one of which is expressed (the others being referred to as 'silent'). One well-known example is the mating type genes of the yeast *Saccharomyces cerevisiae* where the alteration of the mating type is associated with the replacement of the gene at the expression site by the alternative version. All strains, however, continue to possess both types of gene at the silent loci (see Sprague *et al.*, 1983, for details). A more complex example is that of the antigenic variation that occurs with the trypanosome *T. brucei* (reviewed by Boothroyd, 1985). This organism carries a large number of silent genes for the surface antigen. A change in the nature of the expressed gene can be accomplished by movement of a gene to an expression site, or by the movement of expression signals to a position adjacent to a previously silent gene. Although variation of this type is not at present likely to be directly exploitable in strain improvement programmes, knowledge gained from the study of such systems has proved useful in elucidating the mechanisms involved in gene expression and stability.

3.2.2 *Recombination*

In agriculture, 'strain improvement programmes' have in effect been carried on for thousands of years without any formal knowledge of genetics. Much of this has been by means of selection of spontaneous variants, but it has also been recognized for a very long time that a cross between two strains with different desirable characteristics is a good way of achieving still further improvements, not only in the combination of those characteristics, but also in the somewhat ill-defined property of 'hybrid vigour'. This is normally ascribed to the elimination of the deleterious effects of recessive mutations that accumulate in highly inbred strains. Although it is risky to extend the analogy to microorganisms, it is certainly true that in many cases a strain subjected to a series of mutation and selection steps, while showing an improvement in the yield of the specific product, will at the same time suffer a deterioration in other properties such as its growth characteristics or ability to sporulate. Production of recombinants between two high-yielding strains might therefore be a way of restoring some of the vigour of the wild type strains, as well as achieving a further increase in yield. Although, as will be discussed subsequently, the results have often been disappointing, it is important to understand the ways in which such recombinants can be produced.

Firstly, many fungi exhibit a true sexual process which consists (in a very generalized form) of fusion of two haploid parental cells to form a *heterokaryon*. This contains nuclei of both parental types in a common cytoplasm. These nuclei still exist separately, so recombination does not occur at this stage; complementation, however, can be demonstrated in a heterokaryon. The duration of the heterokaryotic state varies from ephemeral to indefinite (depending on the species) and it is succeeded by fusion of two nuclei to give a true diploid cell, usually within a specialized structure. In the yeasts, this diploid state may persist, but in the filamentous fungi the formation of the diploid nucleus is normally followed by meiosis with the production of haploid spores. Genetically, this process is the same as that which occurs in higher organisms, with the important exception that for plants and animals it is the haploid state and not the diploid that is the specialized, transitory form. Fungal genetics therefore has the advantage that the progeny that are examined are haploid, and their genetic composition is therefore more readily determined from the phenotype.

Unfortunately, with a number of commercially important fungi, including *Penicillium chrysogenum*, it has not been possible to demonstrate a sexual cycle. Recombinational analysis and breeding only became possible in such species with the discovery of the *parasexual cycle*. Although, in the heterokaryon, the two nuclear species are normally maintained as such (with nuclear fusion only occurring in specialized structures), it is sometimes possible for fusion of nuclei to occur within the heterokaryotic mycelium, giving rise to a relatively stable diploid form which divides mitotically and can produce asexual diploid spores. The diploid nuclei are not completely stable and may undergo haploidization by a failure of chromosome segregation at cell division (a process known as non-disjunction). Since the loss of chromosomes is random, the haploid segregants will contain a mixture of chromosomes from the two parental strains. The usefulness of the parasexual cycle is increased by the occurrence of mitotic crossing-over, which results in the recombination of genes within pairs of homologous chromosones; this occurs at a much lower frequency than meiotic crossing-over.

Both sexual and parasexual cycles are dependent on the initial formation of a heterokaryon, which generally restricts their application to members of the same species. Even within a species there are commonly barriers to the successful establishment of heterokaryosis. *Protoplast fusion* provides a powerful tool for overcoming this difficulty. Protoplasts can be obtained by the use of specific lytic enzymes or inhibitors of cell wall synthesis in the presence of osmotic stabilizers. Treatment of a mixture of protoplasts with polyethylene glycol (PEG) causes the formation of extensive aggregates. Within these aggregates, there is intimate contact between large areas of membranes of adjacent protoplasts which then appear to fuse, producing cytoplasmic bridges. Transfer to an osmotically stabilized growth medium (with at least partial removal of PEG) results in the regeneration of cell wall and outgrowth of progeny. With filamentous fungi, the progeny are commonly heterokaryotic, and, as with the parasexual cycle, these may give rise to diploid types. With many yeast species, the heterokaryotic stage is transient and cannot be detected, so that the first identifiable products of protoplast fusion are true diploids.

Protoplast fusion techniques can also be applied to bacteria, and in particular have been used very successfully with *Streptomyces* species. Other methods of genetic exchange in bacteria (see below) consist of the transfer of only a part (often a very small part) of the donor chromosome to the recipient cell. In contrast, protoplast fusion results in the formation of a quasi-diploid state containing the complete genome of both partners. Although this state is often only a transient one, the possibilities for recombination are much more extensive than with other techniques.

The technique of protoplast fusion has enormous potential, since it appears that protoplasts of any two species can be fused, irrespective of taxonomic relationships. Interspecies crosses are therefore possible, and in some cases intergeneric crosses may be successful. Across wider taxonomic boundaries, however, although the protoplasts may be seen to have fused, the differences in chromosome organization are too extensive for any viable recombinant progeny to be formed. Transfer of genetic material between such organisms then requires the techniques of *in-vitro* genetic manipulation. One of the most important applications of cell fusions (but outside the scope of this chapter) is in the construction of *hybridomas* (cells that produce *monoclonal antibodies*). For a further discussion of the techniques and potential of protoplast fusion, see Hardy and Oliver (1985).

As well as protoplast fusion, there are three more familiar methods for achieving recombination in bacteria: *transformation*, *transduction* and *conjugation*. Bacterial transformation consists of the uptake of naked DNA by bacteria in a state of competence. For some bacterial species, competent cells are obtained merely by taking a culture at a specific phase of growth, but this is by no means universally applicable. For a long time, transformation was restricted for practical purposes to pneumococci, *Bacillus* species and *Haemophilus influenzae*. However, the development of genetic engineering, in which transformation plays an essential part, has stimulated investigation of transformation systems in other bacteria. It now seems likely that any species of bacterium can take up isolated DNA if the correct conditions can be found. In the case of *E. coli*, competent cells can be obtained by treatment with cold calcium chloride, and DNA uptake is stimulated by a brief heat shock. Transformation of many bacteria, notably *Streptomyces* and *Bacillus* species, can be achieved by a process similar to protoplast fusion: protoplasts will take up plasmid DNA in the presence of PEG. In all cases, the transforming DNA must either be sufficiently homologous to that of the recipient for recombination to occur, or it must form an intact replicon (e.g. plasmid or phage) that can be maintained in the recipient host. Although in some cases heterologous DNA can be taken up, it has no genetic consequences. In those systems that are capable of taking up linear chromosomal DNA, the upper size limit is usually dictated by the practicality of handling large DNA molecules, which renders it impossible to produce anything more than a partial diploid, even of transitory nature.

This applies with even greater force to *transduction*, i.e. phage-mediated gene transfer, where the size of the DNA transmitted is limited by the size of the phage particle. In most cases, this implies a limit of some 1% of the chromosome. Transduction is therefore a powerful tool for fine-structure genetic analysis, and for the transfer of a very limited number of genes without affecting the overall genetic composition of the recipient. Transduction is, however, also limited by the host specificity of the transducing phage, so that it is only available for gene transfer between strains of the same or very closely related species.

Conjugation is the nearest that bacteria get to a true sexual process, but it differs in that the gene transfer is undirectional and unequal. In other words, there is transfer of a copy of part of the DNA of the donor cell into the recipient. Plasmids (integrated or autonomous) play an essential role in this process, and the DNA transferred may in different cases consist of either the plasmid itself or part of the donor chromosome. The whole chromosome is very rarely transferred. Conjugation has now been shown to occur in a wide variety of bacterial genera, not only among the Enterobacteriaceae and other Gram-negative bacteria, but also in many Gram-positive organisms (Thompson, 1986).

3.2.3 Application of manipulations in vivo *to increase enzyme production*

It must be constantly borne in mind, when considering the application of genetic techniques to strain improvement, that this is only one side of the coin. Levels of enzyme production, as well as of production of primary and secondary metabolites, are extremely susceptible to variations in environmental conditions, e.g. growth temperature, pH, aeration, medium composition, and the phase of growth. The effect of growth conditions on production is highly strain-dependent, and therefore needs to be optimized for each strain, including each new variant isolated by the above methods.

The target sites for mutation to increase enzyme production were described in 3.1. However, in order to apply this knowledge, it is necessary to have some *means of selecting or identifying overproducing or constitutive mutants*. If the enzyme confers a selective advantage, this can be exploited to select improved strains: overproduction of catalase causes resistance to hydrogen peroxide; overproduction of beta-lactamase may confer an elevated level of resistance to penicillin. Constitutive mutants may be selected by using substrate analogues which are poor inducers of the enzyme; thus phenyl-beta-galactoside can be used to select constitutive producers of beta-galactosidase. In some cases, specific compounds are available which block induction of specific enzymes, e.g. 2-nitrophenyl beta-fucoside prevents induction of beta-galactosidase by lactose or other inducers. Growth in the presence of the specific substrate plus the induction inhibitor is selective for constitutive mutants. If direct selection for constitutive or overproducing mutants is not possible, the next best thing is a rapid screening method: for example, constitutive mutants of beta-galactosidase can be detected by spraying the plates with a chromogenic, non-inducing substrate.

However, although methods such as these have achieved notable results in the past, genetic manipulation *in vitro* (see below) is a far more powerful approach and should now be regarded as the method of choice for increasing enzyme production.

3.2.4 Overproduction of primary metabolites

Commercially important primary metabolites, such as amino acids, are produced either as the end products of specific metabolic pathways (which are commonly branched) or they may be intermediates in such pathways. The principal factor that affects production of such primary metabolites is the existence of feedback repression and inhibition of an enzyme early in the synthetic pathway. An extremely useful way of selecting mutants which are feedback resistant involves the use of antimetabolites. These are analogues of the end-product of the pathway, which are toxic to the cell because they cause feedback inhibition or repression (and so prevent the formation of the genuine product), but they are unable to substitute for the genuine product in its essential role within the cell. For example, 5-methyltryptophan blocks the production of tryptophan in this way, but cannot be used for protein synthesis. Selection of mutants resistant to 5-methyltryptophan will yield strains which are feedback-resistant, and which will therefore be overproducers of tryptophan. A similar approach has been used for a number of other amino acids and other primary metabolites (Elander, 1987).

Selection of auxotrophic mutants can be used to overcome feedback repression or inhibition effects in the production of related primary metabolites. For example, in Figure 3.2, the initial enzyme 1 is subject to concerted feedback inhibition by the two end-products C (the desired compound) and E (an essential metabolite). In this case, the removal of enzyme 3 will not only divert more of the branch-point compound B into

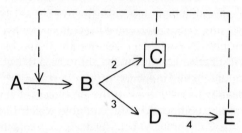

Figure 3.2 Concerted feedback inhibition/repression by the end products of a branched pathway. ──→ feedback inhibition/repression.

production of C, but will also prevent formation of E which is then supplied in limiting amounts to overcome the concerted feedback effect. This is actually a simplified representation of the lysine-producing pathway of *Corynebacterium glutamicum*, where enzyme 1 (aspartokinase) is controlled by feedback inhibition requiring the simultaneous presence of both lysine (C) and threonine (E). Threonine auxotrophs fed with low-level threonine produce over 40 g of lysine per litre (Nakayama *et al.*, 1966). For further examples, see Elander (1987).

3.2.5 Overproduction of secondary metabolites

In general, the control of secondary metabolite production is not so well understood. Although control mechanisms of the same nature as those involved in the control of primary metabolism (with possibly others as yet uncharacterized) are likely to operate, the interactions are so much more complicated that it is usually not possible to predict the nature of the mutations to be looked for. Since it is thus not possible to design a rational improvement programme, the main thrust has been, and still is in many cases, a totally empirical approach. This consists of repeated cycles of screening colonies of mutagen-treated cultures and selection of variants which produce elevated levels of the required secondary metabolite. This can be best illustrated by reference to the most familiar example, the production of penicillin by *Penicillium chrysogenum* (section 2.5.1).

Further information on the control and genetic improvement of production of penicillin and other antibiotics can be found in numerous reviews (for example Hardy and Oliver, 1985; Elanderm, 1987).

Most, if not all, of the commercially important strains of *P. chrysogenum* are descended from a single strain, NRRL 1951, which was isolated in 1943. In tracing the development of this strain it is important to remember that the level of production of penicillin, as with other secondary metabolites, is extremely susceptible to variations in a wide range of growth conditions. The figures obtained by different laboratories may therefore differ quite considerably for any one strain. Wherever possible, therefore, the comparisons drawn below are taken from a single source. Strain NRRL 1951 produced about 100 units of penicillin per ml in submerged culture. From a large number of spontaneously occurring variants a strain (NRRL 1951.B25) was selected which gave a yield of 250 units per ml, i.e. an increase of two- to threefold. An extensive search for further spontaneous variants of B25 failed to produce any higher-yielding strain. However, x-ray mutagenesis of B25 was more successful, resulting in the isolation of strain X-1612 with a penicillin yield of 500 units per ml, twice that of B25 and five times that of the original strain NRRL 1951. Strain X-1612 was the progenitor of the 'Wisconsin family' of strains of *P. chrysogenum*.

Ultraviolet irradiation of strain X-1612 produced, among a number of colonies showing unaltered or decreased penicillin production, one strain (Q176) with a markedly enhanced yield of penicillin (900 units per ml). This was isolated in 1945, and soon became widely used, not only as a production strain but also for other strain development programmes. These early advances are summarized in Figure 3.3.

All the strains up to this point, like the wild-type ancestor, produced the yellow pigment chrysogenin, which contaminated the penicillin and had to be extracted. This not only increased the cost of production but also resulted in the loss of some of the antibiotic. However, in 1947, a variant (BL3-D10), which was completely devoid of pigment, arose from a further ultraviolet irradiation of Q176. Although the production of penicillin by this strain was somewhat lower than that of Q176, the absence of pigment made this strain

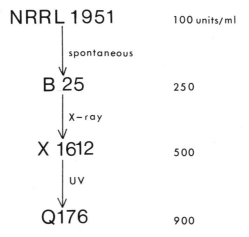

Figure 3.3 Early stages in the development of penicillin production by *P. chrysogenum*.

sufficiently attractive to be used as the breeding stock for further development, thus illustrating the point that characteristics other than product yield have commercial significance. Further developments occurred by a series of chemical mutagenesis steps, culminating in strain 51–20 which produced 2500 units per ml. Most industrial production strains currently used are derived from 51–20, with production levels many times higher. For a more detailed consideration of the history of penicillin production strains, see Elander (1987).

The next point that needs to be considered is the strategy to be adopted for screening and selecting improved strains. In the early stages of a strain improvement programme, there will be many factors that interact to limit the production of the desired metabolite. Elimination of any one of these control processes will therefore give a substantial increase in yield, so it will not be too difficult to isolate improved strains by a totally blind procedure. In essence, this involves picking individual colonies from the mutagenized culture, growing each of them in liquid culture, and determining the yield of product. However, as the programme progresses, and yields become higher, more and more of these sites are 'used up'. Therefore, the percentage of mutants with improved yields is expected to decrease, and also the percentage increase that is likely to be obtained at each stage becomes less.

The first difficulty may be partly overcome by using a rapid screening method that can be directly applied to the colonies on the isolation plate, for example by spraying the plate with a suspension of a sensitive indicator bacterium. Colonies that show a greater ratio of inhibition zone to colony diameter can then be picked and tested in a flask fermentation. Other rapid screening methods are available (Elander, 1987). These methods all depend on the assumption that there is some degree of correlation between performance on a solid medium and that attained in submerged culture. While this correlation is by no means perfect, the assumption has sufficient validity to enable many of the isolates with gross reductions in yield to be discarded, so enabling more of the potentially useful strains to be tested. A widely used alternative is to adopt an automated screening procedure which enables the levels of antibiotic to be determined in very large numbers of liquid cultures.

The second difficulty is that the percentage increases that are likely to be obtained become smaller as the yield is increased. For example, the first step in the improvement of penicillin production by *P. chrysogenum* produced an increase in titre from 100 to 250 units per ml, i.e., an increase of more than 100%. If we are looking for improvements to a strain that already produces say 10 000 units per ml, after a long history of strain improvement, it is unrealistic to expect further improvement of 100% (to 20 000 units per ml), but an increase of 10% (to 11 000 units per ml) would be very useful. The problem then is to distinguish a genetic improvement of this magnitude, from batch-to-batch variation. Although this variation can be reduced by careful control of growth conditions, it can never be entirely eliminated. Furthermore, the optimum conditions for the mutant may be different from those for the parent strain. One answer might appear to be to test each isolate in two or three replicates; this, however, would reduce the number of such isolates that could be tested, thereby reducing the chances of identifying improved mutants which occur infrequently. Alikhanian (1962) considered a mathematical approach to this problem and concluded that in most cases the most effective solution would be to test the greatest possible number of isolates once only. A limited number of these isolates would have to be re-tested, with replicates, and if necessary, more cycles of re-testing—each time using more replicates with fewer isolates—until those variants with the best consistent performance can be identified. It may then be advantageous to carry out the next cycle of mutagenesis and selection on two or three of the best variants rather than singling out one only for further treatment. See Rowlands (1984*a*,*b*) and Elander (1987) for a more detailed consideration of strategies and techniques for strain improvement programmes.

It is often possible to obtain variants with an altered capability for production of specific secondary metabolites by looking for other characteristics. Thus many useful overproducers of penicillin were originally isolated because of their altered colonial morphology. Auxotrophic mutants have also been important in this respect. The production of secondary metabolites is very sensitive to auxotrophic mutation; in the case of penicillin, this effect seems usually to operate in the wrong direction, i.e. the auxotrophs are usually poor penicillin producers. In some cases, however, the auxotrophic mutation is associated with enhanced antibiotic production. In the Merck series of mutants of the streptomycin-producing organism *Streptomyces griseus*, an increase in titre from $1\,mg\,ml^{-1}$ to $2\,mg\,ml^{-1}$ was associated with loss of the ability to make vitamin B_{12} (Demain, 1973). Auxotrophic mutations of tetracycline-producing organisms have also been found to give enhanced yields of antibiotic. These effects may be due to the existence of a branched pathway (see Figure 3.1) so that abolition of synthesis of a primary metabolite which is also an essential compound for growth results in the diversion of an intermediate into secondary metabolite synthesis (see Demain, 1973 for further discussion), or may simply be due to two unrelated mutations.

Although many antibiotics are inhibitory to the organism that produces them, this does not directly affect production, since growth is usually complete before antibiotic production occurs. Nevertheless, selection for resistance to that antibiotic may result in enhanced levels of production, or vice versa. For example, mutants of *Streptomyces noursei* producing high levels of nystatin were resistant to 20 000 units per ml, whereas the parent strain was inhibited by 2000 units per ml, and a non-producing mutant was sensitive to 20 units per ml (Dolezilova *et al.*, 1965).

As discussed earlier, the discovery of the parasexual cycle and its applicability to *P. chrysogenum* stimulated interest in the breeding of improved strains, and the elimination of undesirable characteristics, by crossing strains from different lines leading to the production of heterokaryons, diploids or haploid recombinants. From a cross of two penicillin-producing strains, Sermonti (1959) was able to isolate a diploid which was more productive than either parent, although in other crosses neither the diploids nor the recombinant haploids were more productive than the parents. Alikhanian (1962) also showed that in most cases the diploids and the haploid progeny produced less penicillin than either parent, although one improved diploid was obtained. Many other workers have described results which are disappointing in terms of penicillin yield, although much valuable information has come from such studies (see reviews by Ball, 1973, and MacDonald and Holt, 1976). There are several possible reasons for this difficulty. One is that many of the mutations involved in increases in penicillin yield may be recessive, in which case the diploid would show a decreased yield. Secondly, in order to obtain diploids, the strains involved usually have to be 'marked' genetically by the introduction of auxotrophic mutations or by the use of spore colour mutants, and this often results in a diminished yield of antibiotic. Thirdly, when strains of different lineage are used, the progeny are often found to contain the genome of one or the other parent, with no recombinant progeny being obtained. This is most probably because the extensive series of mutations to which these strains have been subjected has resulted in chromosome rearrangements which restrict or prevent the emergence of recombinants.

Even if the diploid itself does not show any advantages over the haploid parents, either in antibiotic yield or in growth characteristics, it may still provide useful material for further cycles of mutation and selection, particularly if the parental strains have become refractory to mutation. Alikhanian (1962) found that uv irradiation of a heterozygous diploid yielded many more variants with enhanced penicillin production than did equivalent treatment of either parental haploid.

All of the attempts at breeding referred to up to this point have been essentially empirical; crosses have been made between strains without any knowledge of the genetic nature of their titre-increasing mutations. This is virtually inevitable in the case of *P. chrysogenum* because of the rudimentary state of knowledge of the genetics of this organism, although Ball (1973), for example, has described attempts at planned breeding of *P. chrysogenum*. With the discovery that some strains of *Aspergillus nidulans* (in which formal genetics is very much further advanced) are able to produce penicillin, a more thorough study of the genetic factors affecting penicillin yield became possible (see MacDonald and Holt, 1976). Although this is of itself not of direct practical benefit in biotechnology, fundamental knowledge of this nature may eventually permit the by-passing of the law of diminishing returns that besets the empirical approach.

This section has drawn almost exclusively on penicillin production by *P. chrysogenum* as an example to illustrate the application of genetic techniques to obtain increases in production of secondary metabolites. The principles can be extended to other fungi and, selectively, to actinomycetes, especially with regard to antibiotic production.

3.2.6 *Production of novel metabolites.*

Although the most dramatic developments in this field have come from *in-vitro* methods (see 3.3), there are some examples of mutations which alter the nature of the final product. Wild-type strains of *Streptomyces aureofaciens* produce mainly chlortetracycline, but mutants can be found which accumulate tetracycline instead. Other mutants have been isolated which produce different tetracycline derivatives due to blocks at different stages of the synthetic pathway (McCormick *et al.*, 1957, 1958). In the case of aminoglycoside antibiotics, a technique known as 'mutational biosynthesis' has been used (Shier *et al.*, 1969; Nagoka and Demain, 1975). Mutants blocked in the synthesis of the deoxystreptamine or streptidine moiety are fed with analogues of that part of the molecule, which may be incorporated by the organisms to give rise to a novel antibiotic. Further examples of applications of this approach are described by Demain (1973) and Elander (1987).

3.3 Manipulations *in vitro*

The importance of *in-vitro* genetic manipulation (alternatively known as DNA cloning, gene cloning, recombinant DNA technology, or—somewhat emotively and inaccurately—genetic engineering) is, firstly, that it gives us the ability to by-pass the multiplicity of mechanisms that restrict gene transfer between unrelated organisms, and, secondly, that the highly precise nature of the recombinant process enables extremely accurate and subtle manipulations to be performed. These techniques depend on cutting DNA into specific fragments using enzymes known as restriction endonucleases, and joining (ligating) the fragments using another enzyme, DNA ligase. In this way, foreign DNA fragments can be inserted into a vector, i.e. a DNA molecule which is capable of being maintained and replicated by the host cell. The fundamental features of genetic manipulation are described below; for a fuller treatment of the subject, see Old and Primrose (1985) or Brown (1986).

Table 3.1 Examples of restriction endonucleases.

Enzyme	Recognition and cleavage site	Source of enzyme
(a) 6 base recognition site; 5′ sticky ends		
Eco R1	G↓AATTC	Escherichia coli
Hind III	A↓AGCTT	Haemophilus influenzae
Bam H1	G↓GATCC	Bacillus amyloliquefaciens
Sal G1	G↓TCGAC	Streptomyces albus
(b) 6 base recognition site; 3′ sticky ends		
Pst 1	CTGCA↓G	Providencia stuartii
(c) 6 base recognition site; flush ends		
Sma 1	CCC↓GGG	Serratia marcescens
Hpa 1	GTT↓AAC	Haemophilus parainfluenzae
(d) 4 base recognition site; 5′ sticky ends		
Sau 3A1	↓GATC	Staphylococcus aureus
Mbo 1	↓GATC	Moraxella bovis

3.3.1 Basic techniques of in-vitro genetic manipulation

3.3.1.1 Cutting and joining DNA.
Restriction enzymes are so called because they restrict the successful uptake of foreign DNA by a bacterial cell. These enzymes are highly specific in the DNA sequence that they recognize; the most useful type are those which cleave the DNA within the recognition site, which most commonly consists of a sequence of four or six base-pairs. A very large number of restriction endonucleases has now been characterized, and a few examples are listed in Table 3.1. The recognition site is symmetrical, or palindromic, in the sense that the sequence on the complementary strand (read from right to left because of the opposite polarity of the two strands) is identical to that shown in the table. This means that an enzyme such as the *Eco* R1 endonuclease, which cuts asymmetrically within the recognition site, produces DNA fragments with a four-base single-strand extension at the 5′ end (see Figure 3.4). These single-strand extensions are cohesive ('sticky') due to the possibility of hydrogen bond formation between them. Hydrogen bonds between four base-pairs are not strong enough to hold the fragments together at 37°C, but the stability is greater at lower temperatures. The gaps can then be resealed (i.e. the sugar-phosphate backbone is re-formed) by DNA ligase.

The 3′ sticky ends formed by enzymes such as *Pst* 1 can be ligated in the same way. DNA fragments without sticky ends (blunt ended) such as those produced by *Sma* 1 can also be joined together, but with greatly reduced efficiency.

3.3.1.2 Vectors.
Since restriction enzymes and DNA ligase will act in the same way on any DNA molecule irrespective of its source, they provide a

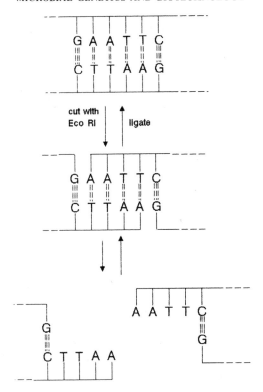

Figure 3.4 Action of restriction endonucleases and DNA ligase.

method of joining heterologous DNA fragments. In particular, they make it possible to incorporate pieces of foreign DNA into a vector DNA molecule which is capable of replication within the chosen organism. Without being so incorporated, the foreign DNA would not be replicated or inherited.

Some features of one of the best-known plasmid vectors, pBR322, are shown in Figure 3.5. It carries determinants for resistance to ampicillin and tetracycline, which enable the selection and identification of clones carrying the plasmid. Figure 3.5 also shows that pBR322 has unique sites for a variety of restriction endonucleases which are available for insertion of DNA fragments. Insertion of a *Pst* 1 fragment at the *Pst* 1 site, which is within the ampicillin resistance gene, will usually result in destroying the function of that gene so that a functional beta-lactamase is no longer made. Recombinant plasmids can therefore be detected by their ability to confer resistance to tetracycline but not to ampicillin. Similarly, insertion of DNA at the *Bam* H1 or *Sal* G1 sites (and in some cases at the *Hind* III site) results in inactivation of the tetracycline resistance gene. Insertion of an *Eco* RI fragment, however,

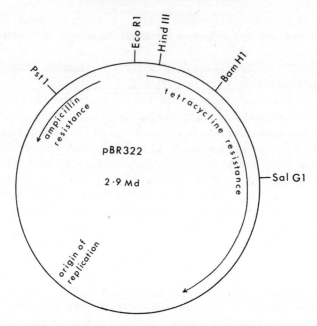

Figure 3.5 Principal features of the structure of pBR322.

does not result in insertional inactivation. There is a wide variety of other plasmid vectors available with different arrangements of unique restriction sites which are therefore suitable for cloning DNA fragments produced by other restriction enzymes.

Most of these vectors are multi-copy plasmids, that is, a cell will normally carry more than one copy, commonly between 10 and 50, but in some cases many more. Furthermore, replication of these plasmids is not affected by the inhibition of protein synthesis by chloramphenicol. This treatment, however, prevents the initiation of new rounds of chromosome replication; chloramphenicol therefore causes a considerable degree of amplification of the plasmid relative to the chromosome, which makes it very much easier to isolate plasmid DNA.

Bacteriophages can also be used as vectors; most of these are based on the temperate *E. coli* phage lambda. Since the amount of DNA that can be packaged into a phage head is subject to certain limits (betweeen 80% and 105% of the normal size) the development of useful vectors involves removing up to 20% of the DNA from the 'non-essential' regions to make room for the insertion of the DNA fragment to be cloned. These phages are then known as insertion vectors. An alternative approach involves so-called 'replacement vectors', which contain two sites for the appropriate restriction enzyme, on either side of a non-essential region of DNA. Cloning is then carried out by replacing that region with the required fragment of foreign DNA. Replacement vectors allow the cloning of much larger DNA fragments, up to about 24 kb. For further information on the use of lambda vectors for gene cloning, see Dale (1987).

3.3.1.3 *Use of hosts other than* E. coli. The vectors referred to so far are only suitable for cloning in *E. coli*. However, *E. coli* is far from ideal for many commercial purposes. Very few proteins are secreted by *E. coli*; extraction and purification of proteins from cells on a commercial scale is much more difficult (and hence costly) than when the proteins are found in the culture supernatant. The accumulation of the protein within the cytoplasm also tends to limit the achievable production level, as the protein is likely to precipitate at the very high concentrations attained. A second disadvantage of *E. coli* is that the lipopolysaccharide (LPS) of the *E. coli* cell envelope has toxic effects, which necessitates extensive purification and quality control checks on products for therapeutic use. Other bacterial hosts, such as *Bacillus subtilis*, which naturally secretes a variety of proteins and does not contain the toxic LPS cell wall component, have been used in an attempt to overcome these two problems (Hardy, 1985). *Streptomyces* species are also widely used for gene cloning, but primarily for the specific purpose of investigation and manipulation of antibiotic biosynthesis (Hopwood *et al.*, 1983). However, these other bacterial systems are much less developed than *E. coli*, which is still the organism of choice for initial studies of specific genes.

A third problem concerns the expression and recovery of biologically active products. Many genes, especially from higher organisms, are not normally expressed when inserted into *E. coli*. Although this can usually be overcome by the use of specially designed *expression vectors* (see below), or in some cases by the use of an alternative prokaryotic host, the protein produced may still not be in a natural, biologically active state. This may be due to a failure to take up the correct secondary/tertiary structure, which may lead to precipitation or degradation. Alternatively, problems may be caused by the absence of any one of a large variety of mechanisms of post-translational modification (such as glycosylation, cleavage of specific peptides, or modification of certain amino acid residues). One way of attempting to overcome this problem is to use a eukaryotic host. The simplest such host is the yeast *Saccharomyces cerevisiae*, which represents a considerable improvement in the expression of biologically active products of genes of higher organisms (Rothstein, 1985). This is still not without problems; for example the glycosylation carried out by *S. cerevisiae* is not identical to that carried out in human cells.

The current trend therefore is towards the development of systems for expressing such genes in animal cell lines and of techniques for growing such manipulated cells in large-scale culture to obtain the desired product. (Spier, 1987). Systems are also under development for the genetic manipulation of plant cells, both for product formation in culture and for the production of plants with novel properties (Fowler, 1987).

Further discussion in this chapter will be mainly confined to the use of *E. coli*, since this is the simplest and most versatile system with which to illustrate the methods used.

3.3.1.4 *Identification of recombinants.* Having ligated the foreign DNA fragments and the vector, the next step is to re-insert the DNA into the host bacterium by transformation of competent cells. With lambda vectors, a more efficient method is by *in-vitro* packaging of the DNA into phage particles which can then infect the host cells (see Dale, 1987, for further details). Transformed cells can be selected by means of the antibiotic resistance marker of the plasmid vector; it is then necessary to identify those clones that carry recombinant plasmids which include the desired gene. Insertional inactivation, as described above, can establish the presence of a recombinant plasmid, but without determining the nature of the inserted DNA. If the gene is expressed in *E. coli*, it may be possible to identify its presence by screening colonies for novel enzyme production or for the presence of a specific antigen (detected by the use of antibodies). If, as is often the case, the cloned DNA is not expected to lead to any recognizable change in the phenotype of the host, then an alternative approach has to be used.

The most important of these possible approaches involves the use of *DNA probes*. In a DNA molecule, the two complementary strands are held together by hydrogen bonding between the bases on the two strands. If two similar DNA species are mixed together, and the hydrogen bonds broken (for example by raising the temperature), the strands will separate. When the temperature is lowered again, the hydrogen bonds will re-form (i.e., the DNA will re-anneal). Since the two DNA species are similar, some of the resulting double-stranded DNA molecules will be hybrids, that is, they will consist of one strand of one type and one strand of the other. Note that the two strands do not have to be identical, but the strength of the association will depend on the degree of similarity between the two strands.

In order to apply this concept of DNA hybridization to the identification of the presence of a specific gene in a recombinant bacterium, it is necessary to have access to a related DNA molecule which can then serve as the *probe*. The procedure is illustrated in Figure 3.6. The transformed colonies are replica plated to a nitrocellulose filter and lysed to release the DNA; this is denatured and fixed to the nitrocellulose so as to produce a DNA print corresponding exactly to the position of the colonies on the original plate. The DNA print is then hybridized with the probe (which has previously been radioactively labelled); after washing off unhybridized DNA, the position of the radioactive

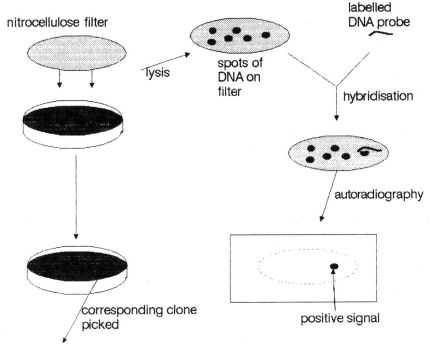

Figure 3.6 Use of a DNA probe to detect recombinant clones.

spots on the filter is revealed by autoradiography to identify the presence of the required DNA. The corresponding clones can then be recovered from the original plate. Other applications of DNA probes are discussed below (see 3.3.2.3).

3.3.1.5 *Strategies of genetic manipulation* in vitro. The easiest approach is to start with the total DNA of the organism of interest, and to fragment it in a random ('shotgun') fashion to give pieces that are of a suitable size for cloning. Ligation of this mixture with an appropriate vector followed by transformation of the host bacterium will give rise to a large number of colonies each carrying a different piece of DNA. This collection of clones constitutes a *gene library* (or more specifically, a *genomic* library, to contrast with a *cDNA library*—see below). If enough clones are obtained, there is a high probability that any specific segment of DNA from the original organism will be present in at least one of the clones. The number of clones needed is a function of the size of the original genome and of the average size of fragment cloned; the greater the genome size or the smaller the cloned fragments, the larger the number of clones needed to constitute a complete library. For example, the human genome consists of some 3×10^9 base-pairs;

if the average insert size is 15 kb (15 000 base-pairs), then in order to have a 99% probability of recovering a specific sequence one would require a library of approximately one million clones.

The random fragments of DNA can be generated in different ways. One possibility is to use restriction enzymes. A restriction enzyme specific for a six-base sequence of DNA would be expected to cut once in every 4096 base-pairs, if the distribution of bases was random. However, restriction sites are rarely distributed in a random fashion, so that the fragments generated will be quite different in size. This will lead to some regions of DNA being under-represented in the library. Although there are ways of improving the randomness of restriction fragments, a better approach is to use truly random fragments of DNA, generated by mechanical shearing.

The size of a complete genomic library can make it difficult (or at least tedious) to locate a specific gene. In addition, eukaryotic genes often contain 'introns', sequences within the gene which are transcribed into a precursor mRNA but are subsequently removed before translation. This means that a complete gene may be very large indeed (often too large to be easily cloned intact). Furthermore, since this processing of the mRNA does not normally occur in bacteria, the genomic clone is unlikely to be of much use in obtaining expression of the gene. The way around these obstacles is to use a *cDNA library*. This involves isolating mRNA and synthesizing the complementary DNA (cDNA) using the enzyme reverse transcriptase. This cDNA can then be cloned as before, usually after the addition of *linkers*. These are synthetic, double-stranded oligunucleotides of defined sequence, containing one or more restriction sites. For example, a linker molecule with the sequence CCGAATTCGG (the other strand is of course complementary to this sequence) contains an *Eco* RI site. These linker molecules are joined to the ends of the DNA to be cloned; digestion with the appropriate restriction enzyme then creates sticky ends which facilitate insertion into the corresponding site on the vector.

It is important to note that whereas a genomic library will be the same (more or less) for a given organism, whatever the starting material, a cDNA library is tissue-specific. A human kidney cDNA library will be different from a human liver cDNA library, since it will reflect the different genes expressed in the different tissues.

Sometimes it is possible to take advantage of this fact to simplify the task of looking for a particular gene. In some cases, cells exist which produce large amounts of a single protein and very little else. Messenger RNA isolated from such cells would therefore be very highly specific for that particular gene. Alternatively, if an enzyme is only produced in response to an external stimulus (in cells grown in culture), then labelling the RNA at the time of induction enables the identification and isolation of the size band of RNA containing that produced from the gene concerned. Although by no means pure, this mRNA will then be highly enriched for the sequences derived from that gene.

3.3.1.6 *Expression.* The final hurdle is to achieve expression of the cloned gene. If bacterial genes are cloned intact, with their promoter sequences, into a closely related species, they may be expressed without any further manipulations. The wider the taxonomic gap, the less reasonable is this expectation. The existence of introns in eukaryotic DNA (see above) exacerbates the problem. The cDNA cloning route overcomes this obstacle, but the DNA is then without a promoter sequence. Specific vectors known as expression vectors have been developed which facilitate the expression of heterologous DNA. In these vectors, the foreign DNA is inserted into a site which is adjacent to a promoter on the vector itself; transcription of the cloned gene can then occur under the direction of the promoter supplied by the vector.

Strategies to maximize expression of cloned genes will be considered further in the next section, as well as in the subsequent descriptions of specific examples.

3.3.2 *Uses of DNA cloning*

3.3.2.1 *Increased enzyme production.* An increase in the yield of a specific enzyme can be obtained in a number of ways. First of all, as mentioned before, most of the commonly used plasmid vectors are multicopy plasmids. Therefore, a gene cloned into one of these plasmids will also be present in the cell as a number of copies, and an enhanced yield of the product of that gene may be expected. An increase in gene copy number can also be obtained by using a lambda phage vector, since after induction of the phage (i.e. commencement of a lytic growth cycle) extensive phage DNA replication occurs. However, in most cases it is more convenient to use plasmid vectors rather than bacteriophages for obtaining enhanced gene expression.

A further increase in enzyme production can be obtained by the use of expression vectors (see above) in which the cloned DNA fragment is placed under the control of an adjacent high-level promoter. In some cases the cloned gene can be expressed so efficiently that the protein produced represents up to 50% of the total cell protein. The drawback is that, since this is non-productive effort from the cell's point of view, such a cell will grow very poorly. Therefore the fermentation process will be inefficient, and in addition any cells that have lost the plasmid, or have stopped making the product for some other reason, will rapidly outgrow the producing cells. Although under laboratory conditions it is the level of production that is the parameter of interest, for commercial purposes the stability of the strain in large scale culture is an equally important consideration.

One way of overcoming the instability of overproducing recombinant strains is to build into the system a mechanism for controlling the expression of the cloned gene. The aim is to grow the culture initially with the gene switched off and then, when a suitable cell density is reached, to change the culture conditions in such a way as to allow expression to start. Under laboratory conditions, this is commonly done by using the *lac* promoter as the expression signal. With this promoter on the vector, and the cloned gene adjacent to it and under its control, expression will normally be switched off due to the action of the *lac* repressor protein. When expression is required, the addition of an inducer such as IPTG (iso-propyl-thiogalactoside) will inactivate the repressor, resulting in a sudden onset of expression of the cloned gene. Since no further growth is required, there is no opportunity for plasmid-free cells to overgrow the producing bacteria. However, IPTG is not cheap, and this strategy would make a commercial scale fermentation prohibitively expensive (apart from the problem of removing and disposing of the IPTG).

An alternative strategy involves the use of the P_L promoter of lambda. This is a high-level promoter which is normally repressed by the action of the cI repressor protein of the bacteriophage. Mutations of the cI gene (cIts) are available that give rise to a temperature-sensitive repressor. An expression vector that carries a cIts gene and in which the cloned gene is under the control of a P_L promoter will result in the cloned gene being repressed when the culture is grown at a reduced temperature (30°); the culture will therefore grow normally until a high cell density is reached, at which point the temperature can be raised, e.g. to 40°. The repressor protein will be inactivated and the required gene product expressed.

3.3.2.2 *Novel products.* The second use of DNA cloning is the one that has attracted most publicity, namely the transfer to an easily-grown microorganism of the ability to make a specific biological product. In this way, hormones for therapeutic use, for example, can be produced from a microbial culture instead of being extracted from human or animal tissues. Some examples (insulin, human growth hormone, interferon and viral vaccines) will be considered in detail in section 3.3.3.

Although these achievements have been dramatic, there are limitations to the immediate commercial potential, since only simple gene products (peptides or proteins) are readily amenable to gene cloning. Furthermore, from a commercial viewpoint, two additional criteria must be met: the existing source of supply must be very expensive or inadequate, and the demand for the product must be sufficient to enable the very considerable research and development costs to be recovered.

The techniques are not limited to the cloning and exploitation of naturally occurring genes. Once the DNA sequence of a gene is known, it becomes possible to design variants of that gene which may specify the production of a protein with new and desirable properties. For example, the exploitation of many enzymes is limited by their thermal instability. If two amino acids that are adjacent in the three-dimensional structure (although in different regions of the amino-acid sequence) are changed to cysteine residues, this will result in the formation of an additional disulphide bridge which leads to a more stable conformation. Other changes could produce an alteration of the structure of the active site resulting in a changed substrate specificity.

It is possible to obtain such alterations by constructing a gene from scratch, using synthetic DNA. This however would be rather a cumbersome way of introducing specific changes at a particular position. A more convenient method involves synthesizing a short piece of DNA containing the desired alteration. This is annealed to a single-stranded version of the original gene, thus providing a primer for the enzymic synthesis of the complete second strand which now contains the required alteration. Cloning of the resulting hybrid DNA will yield a mixture of clones, some containing the original gene and some carrying the altered version.

A major limitation on the application of this technique of *in-vitro* mutagenesis (also known as site-directed mutagenesis) is that a rational approach requires detailed knowledge of both the three-dimensional structure of the protein and of the functional importance of the relevant parts of that structure. Sufficiently detailed structures from x-ray crystallography are available for only a limited number of proteins, and even in these cases it is often not possible to make more than an informed guess as to the functional consequences of altering particular amino acid residues. At present, therefore, these studies are likely to be more valuable in providing us with that sort of information than in yielding commercially valuable products. For more information on this topic, see Fersht (1985).

Many important biological products are not peptides or proteins, but are the end-product of a series of enzymic reactions. If each step is known, then it is possible, theoretically, to clone each gene separately. This would be a very laborious undertaking and would be of doubtful benefit in most cases. It is not even certain that a whole pathway could be reassembled in this manner, since there are often very sophisticated interactions between the enzymes in a metabolic pathway, both amongst themselves and also with the remainder of the cell's metabolic apparatus. Furthermore, in many cases of commercial significance, the pathway concerned is not known in sufficient detail, which is an insuperable obstacle to a direct approach. There have, however, been some successes in cloning the genes for a whole pathway where the genes concerned are grouped together on the chromosome, so that a single large DNA fragment can contain the complete set (Malpartida and Hopwood, 1984).

In the case of existing microbial fermentations, such as antibiotic production, an alternative use of genetic manipulation in strain development is to use as the host a strain which already produces the antibiotic and to add genes from another organism which may by-pass a rate-limiting step or improve the growth characteristics of the producing organism. In the following section there is a description of the application of this approach to the ICI single cell protein process. Genetic manipulation of antibiotic producing strains can also be employed to elucidate the nature of the genes and the biosynthetic pathway involved. (Bailey et al., 1984).

3.3.2.3 *Gene probes.* The use of labelled fragments of DNA for detecting the presence of homologous sequences has been an invaluable research tool for many years. The use of gene probes for detecting specific recombinant clones as described above is one example. A second example, from the field of medical research, involves the study of human genetic disorders. Here, a combination of DNA probes with gene cloning and sequencing techniques has led to a detailed knowledge of the nature of the underlying genetic defect for certain inherited diseases. In some of these cases, gene probes can be used to detect the altered gene, thus making it possible to diagnose the defect in a foetus at an early stage, and also to detect the presence of the defective gene in a heterozygous adult who may show no signs of the disease but is capable of passing the defective gene on to any offspring.

Another field in which gene probes are becoming commercially exploitable is in the detection of infectious agents, especially where these are difficult to culture, or exist in a latent form which may not be easy to detect by conventional means. However, there are a number of limitations which currently restrict the routine application of gene probes. The first of these concerns the labelling procedure. Conventionally, DNA probes are labelled by enzymic incorporation of radioactively labelled nucleotides; these are, however, potentially hazardous, and the decay of the radioactive label means that the probe has a very short shelf-life. Non-isotopically labelled probes can be produced, for example by using biotin-substituted nucleotides, but these are generally somewhat less sensitive.

The sensitivity of the techniques is another limitation, especially in the detection of infectious agents. The usual methods require between 10^4 and 10^6 copies of the target DNA sequence before it can be readily detected, whereas an ideal method for an infectious agent should be capable of detecting a single bacterium or virus in a given sample. Methods are currently under development, for example using a technique known as *target amplification* (Saiki et al., 1985, 1986) which are at least in theory capable of achieving this level of sensitivity.

A further application that has received a great deal of publicity is the use of probes for *DNA fingerprinting* (Jeffreys et al., 1985; Gill et al., 1985). This relies on the fact that human DNA contains a large number of copies of very similar short sequences of DNA (minisatellite DNA). The distribution of these sequences is different for each individual, so that when a restriction digest of the DNA is subjected to electrophoresis a characteristic pattern of bands will be obtained. These bands are visualized by transferring the material from the gel to a nitrocellulose filter (Southern blotting) and hybridizing the filter with the specific DNA probe. Although the overall pattern is characteristic of an individual, it is inherited and represents a combination of certain aspects of the pattern seen with each parent. In this way, not only can the origin of a specific DNA sample (for example from a bloodstain at the scene of a crime) be ascertained by comparison with the patterns obtained from potential suspects, but also questions of relationship (e.g. paternity) between individuals can be at least partly clarified.

3.3.2.4 *Other uses.* Although this chapter concentrates on the applications to biotechnology, the most radical impact of gene cloning has been on our understanding of the basic nature of inheritance and gene expression. This is especially true of mammalian genetics (and human genetics in particular) where a few experiments can now achieve what would previously have taken years of patient observation (and often would not have been possible at all). DNA technology (coupled with other techniques, notably the use of monoclonal antibodies) is also playing a crucial role in attempts to understand the nature of a wide range of human diseases from cystic fibrosis to cancer.

3.3.3 *Genes and biotechnology: applications of gene cloning*

3.3.3.1 *Growth hormones.* Human growth hormone (HGH; somatotropin) is used in the treatment of pituitary dwarfism. Until the advent of genetic engineering, the only source of this material was human pituitary glands that were removed at autopsy. The restricted nature of the supply, coupled more recently with concern over possible transmission of latent viral infections by this material, made this an obvious target for gene cloning.

Initial experiments (Seeburg *et al.*, 1977, 1978) with rat growth hormone involved obtaining mRNA from cells which produce the growth hormone and cloning cDNA produced from that mRNA. Rat growth hormone can be produced from pituitary cells grown in culture, but under normal conditions only a small proportion (1–3%) of the mRNA is specific for the growth hormone. This proportion can be raised by induction with thyroid hormones and glucocorticoids, and further enrichment can be obtained by using mRNA from cytoplasmic membranes rather than from whole-cell extracts. Growth hormone mRNA is then the most abundant species, representing about 10% of the mRNA isolated. Expression of the cloned cDNA was obtained by insertion into the Pst 1 site of pBR322 to put it under the control of the β-lactamase gene. The product was a fused protein consisting of the N-terminal part of the β-lactamase joined to the growth hormone itself.

One disadvantage of this procedure is that the primary translation product of secreted proteins such as growth hormone is a precursor protein rather than the mature form, and that the cloned cDNA will also specify the precursor form. In bacteria, this protein will not be processed into the mature form. The use of a chemically synthesized gene can overcome this problem. Although chemical synthesis is feasible for a gene of this size (the protein consists of 191 amino acids) it is still rather too time-consuming to make it the procedure of choice in most cases. Goeddel et al. (1979a) used a combination of the two approaches to obtain direct expression of mature human growth hormone (HGH) by E. coli. They were able to isolate a specific restriction fragment of about 550 base-pairs containing the coding sequence for amino acids 24–191 (COOH terminus) of HGH. A synthetic DNA molecule was constructed corresponding to the first 24 amino acids of mature HGH, plus an ATG initiation codon. These two fragments were ligated and inserted into an expression vector so that the cloned DNA was under the control of a *lac* promoter. The resulting plasmid gave rise to the production of 2 mg of HGH per litre of culture, corresponding to nearly 200 000 molecules of HGH per cell.

3.3.3.2 *Insulin.* Although insulin, a small protein hormone, is composed of two chains, A and B, it is the product of single gene. The initial translation product of insulin mRNA is a single polypeptide, preproinsulin, from which a signal sequence of 23 amino acids at the amino terminus is removed to give the insulin precursor, proinsulin. This folds up, and a peptide sequence (the C peptide) is removed from the middle to separate the A and B chains. Obtaining expression of insulin by bacterial cells is therefore not at all straightforward. Early experiments on rat insulin (Ullrich et al., 1977; Villa-Komaroff et al., 1978) followed the cDNA route, using mRNA from tissues that produce insulin, and in which therefore insulin mRNA is abundant. This cDNA, however, corresponds to the preproinsulin sequence, containing the signal peptide sequence. While it would be comparatively easy to shorten the sequence to that corresponding to proinsulin, the bacterial cells are unable to carry out the subsequent removal of the C peptide to yield mature insulin.

A different approach was therefore used by Goeddel et al. (1979b) for cloning the human insulin gene. Two DNA sequences were produced synthetically, corresponding to the A and B polypeptides, and each of these was inserted into pBR322 together with a DNA fragment carrying most of the β-galactosidase gene plus its control region. The recombinant plasmid pIB1 produced a hybrid β-galactosidase-insulin B chain protein, under the control of the *lac* operator/promoter region. A similar recombinant plasmid pIA1 specified a fused β-galactosidase-A chain protein. A very high level of expression was achieved, as the hybrid proteins represented about 20% of the total cell protein. Furthermore, these hybrid proteins were insoluble and so could be easily separated from the bulk of the remaining protein. The insulin chains could be cleaved from the β-galactosidase by cyanogen bromide, and after sulphonation were reconstituted to form intact, mature insulin.

These and other developments in the cloning and expression of hormone genes in *E. coli* have been reviewed by Primrose (1986).

3.3.3.3 *Interferon*. Many types of human and animal cells produce, as a consequence of exposure to certain viruses or other inducing agents, glycoproteins known as interferons, which have a molecular weight of about 20 000. Interest in interferons first arose from their ability to render cells resistant to virus attack, and subsequently from the possibility that interferons may be useful as anti-cancer agents. However, supplies of interferon were severely limited; for example, two litres of human blood (one of the principal sources) were required to produce 1 μg of purified human leukocyte interferon. This made interferon an obvious target for gene cloning, at the same time as research was being carried out into improving the production of interferon from cultured human fibroblasts, and purification by affinity chromatography using monoclonal antibodies.

Nagata *et al.* (1980) obtained mRNA from human leukocytes induced with Sendai virus and inserted the cDNA from this, after tailing, into the Pst 1 site on pBR322. A number of isolates were identified which contained interferon (IF) DNA, and some of these were found to produce a protein with interferon activity. Although this approach, when used for other genes (such as growth hormone, see 3.3.3.1 above) resulted in the production of a fused protein incorporating part of the β-lactamase, this did not appear to happen with interferon. This may indicate that the interferon was produced from a translation initiation site associated with the cloned IF sequence. The yield was low (about 20 000 units per litre of culture) which corresponds (if the interferon produced has the same specific activity as authentic interferon) to only one to two molecules per cell.

Goeddel *et al.* (1980), using the same initial cloning strategy, obtained a recombinant plasmid with an insert of about 1000 base-pairs, which could be excised and cloned into another vector to place it under the control of a *trp* promoter. This gave a much higher level of expression (about 480 000 units per litre of culture). Knowledge of the DNA sequence of the 1000 base-pair fragment enabled a further series of manipulations to be carried out to obtain expression of mature leukocyte interferon at a high level. This involved removing the sequence coding for the signal precursor peptide and replacing it with a synthetic fragment including a translation initiation ATG codon. Another DNA fragment was then inserted which carried the *E. coli* trp operator/promoter region and ribosome binding site. One clone obtained after these manipulations produced 2.5×10^8 units of interferon per litre of culture. The interferon produced was shown to have biological activity by its ability to protect squirrel monkeys against encephalomyocarditis virus.

Subsequently, Hitzeman et al. (1981) described the cloning of a human leukocyte IF gene in yeast cells, using a dual-purpose vector able to replicate in either E. coli or yeast. The IF gene was joined to a specific yeast promoter which enabled effective expression of mature interferon in yeast (but not in E. coli.)

3.3.3.4 *Vaccines and diagnostic reagents.* The production of vaccines and of immunological reagents used in diagnosis requires a convenient method of obtaining the organism concerned in reasonably large quantities. Most (but by no means all) pathogenic bacteria can be grown in relatively simple culture media; two important exceptions are the causative agents of syphilis and leprosy. Viruses, on the other hand, are not so readily obtained. Even if a tissue culture system is available, this is considerably more expensive and less convenient than a bacterial culture. Furthermore, there are viruses such as hepatitis B which cannot be grown in tissue culture, and for which no convenient animal system is available (hepatitis B normally infects only man and apes). Cloning DNA from such an organism into E. coli could provide a route for large-scale production of specific antigens for both diagnostic purposes and for vaccine production.

Blood from patients infected with hepatitis B virus (HBV) contains particles (Dane particles) which have an envelope containing the hepatitis B surface antigen (HBsAg) and an inner core containing the core antigen (HBcAg). The double-stranded circular DNA (molecular weight $c.\ 2 \times 10^6$) is contained within the core. A third antigen (HBeAg) is found in the plasma of some infected individuals. Hepatitis B vaccine can be produced from the blood of carriers of the virus, but it requires extensive purification and inactivation, and stringent quality control, to ensure that there is no risk of it containing infectious hepatitis B particles, and similarly to avoid the risk of transmitting AIDS, since the high-risk groups for the two diseases are identical.

Burrel et al. (1979) isolated DNA from Dane particles and cloned fragments of this DNA into the *Pst* 1 site of pBR322. Colonies carrying recombinant plasmids were screened for the production of HBV antigens by a solid-phase radioimmunoassay procedure, and a number of colonies were found which gave a positive response with anti-HBc serum, indicating the production of immunologically active core antigen. This appeared to be present as a periplasmic polypeptide fused to the major portion of the β-lactamase specified by pBR322.

As discussed above (3.3.1.3), E. coli may not be the best host to choose for vaccine production because of the risk of contamination by endotoxins. Hardy et al. (1981) have described the production in *Bacillus subtilis* of the core antigen of hepatitis B virus. Subsequently (Valenzuela et al., 1982;

McAleer *et al.*, 1984) the hepatitis B surface antigen gene has been expressed in yeast, and it is from this source that recombinant hepatitis B vaccine is produced.

An alternative approach which shows much promise is to insert antigen genes into derivatives of the vaccinia virus which can then be used as a live vaccine. For example, Smith *et al.* (1983) and Paoletti *et al.*, (1984) constructed vaccinia recombinants containing the hepatitis B surface antigen gene and showed that vaccinated rabbits rapidly produced antibodies to HBsAg. Paoletti *et al.* (1984) also produced vaccinia derivatives carrying the herpes simplex glycoprotein D gene which were capable of protecting mice against challenge with lethal doses of live herpes simplex virus. Several different genes can be introduced into the same virus, thus creating a live vaccine which can confer immunity to several diseases simultaneously. For example, Perkus *et al.* (1985) produced a recombinant vaccinia strain carrying genes from hepatitis B, herpes simplex and influenza viruses and showed that rabbits immunized with this recombinant strain produced antibodies against all three viruses. The envelope gene of the AIDS virus (HIV) has also been expressed by recombinant vaccinia virus (Chakrabarti *et al.*, 1986), and vaccinia strains carrying a gene from the rabies virus have been produced as a possible method for oral vaccination of foxes in order to control the disease among wild animals (Blancou *et al.*, 1986).

Even if the pathogen itself can be grown on a large scale, there can still be advantages (other than cost) in cloning the antigen-specifying genes and using a non-pathogenic bacterium as the producing organism. It avoids the hazards of large-scale production of a potentially dangerous pathogen, and reduces the risk inherent in using the vaccine. For example, attenuated virus may revert to a virulent form unless extensive precautions are taken, and in the case of a killed vaccine, the effectiveness of the killing procedure has to be carefully monitored. The use of a cloned antigen removes these risks. In addition, the specificity of the vaccine can also be increased, since a single pure antigen is used, which would be expected to reduce the occurrence of damaging side-effects such as those seen occasionally with pertussis (whooping cough) immunization. In fact, it may not even be necessary to use a complete protein, since it may be only a specific part of the protein (epitope) that is responsible for the major part of the immunological response. For example, with the malaria parasite, much of the antibody response to the infecting sporozoite stage is directed against the so-called circumsporozoite protein (CSP), and in particular to a region of that protein that consists of many repeats of a four-amino-acid sequence, asparagine–alanine–asparagine–proline. Ballou *et al.* (1987) used an immunogen, prepared by a recombinant DNA route, which consisted of 32 tandem repeats of this four-amino-acid sequence. This material was not only able to elicit antibody production, but also protected human volunteers against challenge with infectious sporozoites. For shorter peptide sequences, it is usually easier (at least on a small scale) to produce them synthetically. Herrington et al. (1987) used a 12-amino-acid synthetic peptide consisting of three repeats of the four-amino-acid sequence. Peptides this short are usually not immunogenic, so it was necessary to link it to a carrier molecule, in this case tetanus toxoid. Again, this material was shown to be capable of protecting human volunteers against challenge with the sporozoite stage of the malaria parasite.

3.3.3.5 *ICI single cell protein* (see also 2.4.6.10). This was one of the earliest applications of gene cloning to achieve commercial significance. The organism used by ICI to produce single cell protein (SCP) from methanol, the pseudomonad referred to as *Methylophilus methylotrophus*, assimilates ammonia via a two stage pathway (dependent on glutamine synthetase and glutamate synthase) which requires an extra molecule of ATP compared to the glutamate dehydrogenase pathway used by *E. coli*. Therefore, by cloning the *E. coli* glutamate dehydrogenase (*gdh*) gene and introducing it into a glutamate synthase-deficient mutant of *M. methylotrophus*, an increase in the efficiency of the process might be obtained. The procedure by which this was achieved involved a combination of *in-vitro* and *in-vivo* manipulations (Windass et al., 1980). Enzyme assays showed that *M. methylotrophus* carrying *gdh* recombinant plasmids did produce functional glutamate dehydrogenase. The modified organism obtained at the end of these manipulations gave a carbon conversion that was 4–7% higher than that of the original strain.

3.3.4 Safety implications

From the earliest days of genetic manipulation *in vitro*, there has been concern about the implications of the technique for the possible creation of strains with enhanced pathogenicity and also because of the environmental impact of 'new' bacteria. Although much of this concern was without real foundation, perhaps due to the emotive connotations of the term 'genetic engineering', there were rational grounds for believing that there was a hazard. For example an *E. coli* strain producing botulinum toxin could conceivably be very dangerous; insertion of a penicillinase gene into *Streptococcus pyogenes* (still universally sensitive to penicillin) could make the treatment of streptococcal infections very much more difficult. In practice, neither of these examples is now believed to be quite as dangerous as was thought at first. Some of the hazards are more conjectural: what would be the effect if an *E. coli* strain producing insulin were to colonize the human intestine? What would be the effect on the nitrogen balance of the environment if the genes for nitrogen fixation were to become widely disseminated? And, one of the original causes of concern, would cloning genes from oncogenic viruses lead to epidemics of cancer caused by *E. coli*?

Because of fears of this nature, an elaborate system of restrictions and regulations was brought into being, to ensure that those experiments which were thought to carry the greatest risk were performed only under conditions which minimized the chance of the recombinant organisms escaping from the laboratory and establishing themselves in the environment. The precautions are of two kinds: physical containment (ranging from simple safety cabinets to elaborate systems of airlocks and effluent sterilization plant) and biological containment, which principally means the use of 'crippled' strains which carry a number of defects rendering them extremely unlikely to be able to survive outside the laboratory.

The severity of the containment required is related to an assessment of the risks associated with a particular experiment. As our knowledge of the parameters involved has increased—the barriers to gene expression and the difficulty that even routine laboratory strains of *E. coli* have in colonizing the gut—so have the precautions been relaxed, especially at the lower end of the danger spectrum. A wide range of manipulations is now permitted under conditions no more stringent than 'good microbiological practice'.

However, a new concern has arisen more recently in that bacteria have been 'engineered' to carry out specific tasks when released into the environment. For example, frost damage to plants is partly associated with ice formation, which can be promoted by the presence of certain bacteria (such as *Pseudomonas syringae*) that catalyse ice nucleation (Green and Warren, 1985). Strains of *Ps. syringae* have been developed which are lacking this ice nucleation ability. Proposals to colonize plants with these bacteria by spraying them on a large scale aroused considerable concern in the USA. The assessment of the potential risk of deliberately releasing bacteria (whether or not they have been subjected to genetic manipulation) has not yet been satisfactorily resolved. (See Jukes, 1986, for one view of some of the arguments; a more comprehensive treatment of the whole field of risk assessment is provided by Fiksel and Covello, 1986.)

3.3.5 *Future prospects*

The number of easily identifiable, dramatic targets such as insulin and interferon is very limited. One further example, which is currently the subject of extensive legal action over patent rights, is tissue plasminogen activator (TPA) which can stimulate lysis of blood clots and therefore has considerable potential in the treatment of thrombosis. Although the production costs of these recombinant materials is comparatively low, the research and development costs, and (for therapeutic products) the cost of meeting the regulations are enormous, and require that the potential market is large and that existing sources of supply are unsatisfactory. (See Primrose, 1986, for further discussion of the potential and limitations of these applications.) Although many enzymes and peptide hormones are the targets of gene cloning projects, this is likely in many cases to be of more value to medical research than to commercial biotechnology. Animal hormones, for use as growth promoters, and enzymes for commercial use are more obvious targets as are improvements in vaccine production (for veterinary use even more than for human purposes). However, the most widespread commercial application in the short term will probably be in the less dramatic area of improvement in the performance of existing industrial microorganisms, as exemplified by the ICI SCP process.

In summary therefore, the microbial geneticist has three main routes available for consideration for a programme of strain development, and the choice depends primarily on the nature of the product and the extent of fundamental knowledge of the processes involved.

(i) *Random mutation and selection.* Although this has an old-fashioned ring to it, it is likely to be the method of choice for improvement in secondary metabolite production for some time to come, mainly because of the present lack of knowledge of the control mechanisms involved.

(ii) *'Rational' mutation programmes,* such as the use of toxic analogues to select mutants resistant to feedback inhibition and repression. This is mainly useful for primary metabolites and it depends on a certain amount of knowledge of how the pathway is controlled. Both (i) and (ii) can also be assisted by recombinational studies in appropriate cases.

(iii) *Gene cloning.* This should now be the automatic first choice approach if the product is a protein or a polypeptide, and is also useful for introducing specific changes in metabolism. For a wide range of other products, it is undoubtedly the technique of the future, the full potential of which has yet to be achieved.

References

Alikhanian, S. I. (1962) *Adv. appl. Microbiol.* **4**, 1–50.
Bailey, C. R., Butler, M. J., Normansell, I. D., Rowlands, R. T. and Winstanley, D. J. (1984) *Bio/Technology* **2**, 808–811.
Ball, C. (1973). *Prog. ind. Microbiol.* **12**, 47–72.
Ballou, W. R., Hoffman, S. L., Sherwood, J. A., Hollingdale, M. R., Neva, F. A., Hockmeyer, W. T., Gordon, D. M., Schneider, I., Wirtz, R. A., Young, J. F., Wasserman, G. F., Reeve, P., Diggs, C. L. and Chulay, J. D. (1987) *Lancet* **i**, 1277–1281.
Bennett, P. (1985) Bacterial transposons. In *Genetics of Bacteria*, eds. J. Scaife, D. Leach, and A. Galizzi, Academic Press, London.
Blancou, J., Kieny, M. P., Lathe, R., Lecocq, J. P., Pastoret, P. P., Soulebot, J. P. and Desmettre, P. (1986) *Nature (London)* **322**, 373–375.
Boothroyd, J. C. (1985) *Ann. Rev. Microbiol.* **39**, 475–502.

Brown, T. A. (1986) *Gene Cloning*, Van Nostrand Reinhold. New York.
Burrell, C. J., Mackay, P., Greenaway, P. J., Hofschneider, P. H. and Murray, K. (1979) *Nature (London)* **279**, 43–47.
Chakrabarti, S., Robert-Guroff, M., Wong-Staal, F., Gallo, R. C. and Moss, B. (1986) *Nature (London)* **320**, 535–540.
Dale, J. W. (1987) Cloning in bacteriophage lambda. In *Techniques in Molecular Biology*, vol. 2, eds. J. M. Walker and W. Gaastra, Croom Helm, London, 159–177.
Demain, A. L. (1973) *Adv. appl. Microbiol.* **16**, 177–202
Dolezilova, L., Spizek, J., Vondracek, M., Paleckova, F. and Vanek, Z. (1965). *J. gen. Microbiol.* **39**, 305–309.
Elander, R. P. (1987) Microbial screening, selection and strain improvement. In *Basic Biotechnology*, eds. J. Bu'Lock and B. Kristiansen, Academic Press, London, 217–251.
Fersht, A. (1985) *Enzyme Structure and Mechanism*. Freeman, San Francisco.
Fiksel, J. and Covello, V. T. (eds.) (1986) *Biotechnology Risk Assessment*. Pergamon, Oxford.
Fowler, M. W. (1987) Products from plant cells. In *Basic Biotechnology*, eds. J. Bu'Lock and B. Kristiansen, Academic Press, London, 525–544.
Galloway, J. L. and Platt, T. (1986). Control of prokaryotic gene expression by transcription termination. In *Regulation of Gene Expression*, eds. I. R. Booth and C. F. Higgins, Cambridge University Press, 155–178.
Gill, P., Jeffreys, A. J. and Werrett, D. J. (1985) *Nature (London)* **318**, 577–579.
Goeddel, D. V., Heyneker, H. L., Hozumi, T., Arentzen, R., Itakura, K., Yansura, D. G., Ross, M. J., Miozzari, G., Crea, R. and Seeburg, P. H. (1979a) *Nature (London)* **281**, 544–548.
Goeddel, D. V., Kleid, D. G., Bolivar, F., Heyneker, H. L., Yansura, D. G., Crea, R., Hirose, T., Kraszewski, A., Itakura, K. and Riggs, A. D. (1979b) *Proc. natl. Acad. Sci. USA* **76**, 106–110.
Goeddel, D. V., Yelverton, E., Ullrich, A., Heyneker, H. L., Miozzari, G., Holmes, W., Seeburg, P. H., Dull, T., May, L., Stebbing, N., Crea, R., Maeda, S., McCandliss, R., Sloma, A., Tabor, J. M., Gross, M., Familletti, P. C. and Pestka, S. (1980) *Nature (London)* **287**, 411–416.
Green, R. L. and Warren, G. J. (1985) *Nature (London)* **317**, 645–648.
Hardy, K. G. (1985) Bacillus cloning methods. In *DNA Cloning*, vol. II, ed. D. M. Glover, IRL Press, Oxford, 1–17.
Hardy, K., Stahl, S. and Kupper, H. (1981) *Nature (London)* **293**, 481–483.
Hardy, K. G. and Oliver, S. G. (1985) Genetics and biotechnology. In *Biotechnology*, eds. I. J. Higgins, D. J. Best and J. Jones, Blackwell Scientific, Oxford, 257–282.
Herrington, D. A., Clyde, D. F., Losonsky, G., Cortesia, M., Murphy, J. R., Davis, J., Baqar, S., Felix, A. M., Heimer, E. P., Gillessen, D., Nardin, E., Nussenzweig, R. S., Nussenzweig, V., Hollingdale, M. R. and Levine, M. M. (1987) *Nature (London)* **328**, 257–259.
Hitzeman, R. A., Hagie, F. E., Levine, H. L., Goeddel, D. V., Ammerer, G. and Hall, B. D. (1981) *Nature (London)* **293**, 717–722.
Hopwood, D. A., Bibb, M. J., Bruton, C. J., Chater, K. F., Feitelson, J. S. and Gill, J. A. (1983) *Trends in Biotechnology* **1**, 42–48.
Jeffreys, A. J., Wilson, V. and Thein, S. L. (1985) *Nature (London)* **314**, 67–73.
Jukes, T. H. (1986) *Nature (London)* **319**, 617.
MacDonald, K. D. and Holt, G. (1976) *Sci. Prog. Oxford* **63**, 547–573.
Malpartida, F. and Hopwood, D. A. (1984) *Nature (London)* **309**, 462–464.
McAleer, W. J., Buynak, E. B., Maigetter, R. Z., Wampler, E., Miller, W. J. and Hilleman, M. R. (1984) *Nature (London)* **307**, 178–180.
McCormick, J. R. D., Sjolander, N. O., Hirsch, U., Jensen, E. R. and Doerschuk, A. P. (1957) *J. Am. Chem. Soc.* **79**, 4561–4563.
McCormick, J. R. D., Miller, P. A., Growich, J. A., Sjolander, N. O. and Doerschuk, A. P. (1958) *J. Am. Chem. Soc.* **80**, 5572–5573.
Nagaoka, K. and Demain, A. L. (1975). *J. Antibiot.* **28**, 627–635.
Nagata, S., Taira, H., Hall, A., Johnsrud, L., Streuli, M., Ecsodi, J., Boll, W., Cantell, K. and Weissmann, C. (1980) *Nature (London)* **284**, 316–320.
Nakayama, K., Tanaka, K., Ogino, H. and Kinoshita, S. (1966) *Agr. Biol. Chem.* **30**, 611–616.
Old, R. W. and Primrose, S. B. (1985) *Principles of Gene Manipulation*, Blackwell Scientific, Oxford.

Paoletti, E., Lipinskas, B. R., Samsonoff, C., Mercer, S. and Panicali, D. (1984) *Proc. natl. Acad. Sci. USA* **81**, 193–197.

Perkus, M. E., Piccini, A., Lipinskas, B. R. and Paoletti, E. (1985). *Science* **229**, 981–984.

Platt, T. (1986) *Ann. Rev. Biochem.* **55**, 339–372.

Primrose, S. B. (1986) *J. appl. Bacteriol* **61**, 99–116.

Rothstein, R. (1985) Cloning in yeast. In *DNA Cloning*, vol. II, ed. D. M. Glover, IRL Press, Oxford, 45–66.

Rowlands, R. T. (1984a) *Enzyme Microb. Technol.* **6**, 3–10.

Rowlands, R. T. (1984b) *Enzyme Microb. Technol.* **6**, 290–300.

Saiki, R. K., Bugawan, T. L., Horn, G. T., Mullis, K. B. and Erlich, H. A. (1986) *Nature (London)* **324**, 163–166.

Saiki, R. K., Scharf, S., Faloona, F., Mullis, K. B. Horn, G. T., Erlich, H. A. and Arnheim, N. (1985) *Science* **230**, 1350–1354.

Seeburg, P. H., Shine, J., Martial, J. A., Baxter, J. D. and Goodman, H. M. (1977). *Nature (London)* **270**, 486–494.

Seeburg, P. H., Shine, J., Martial, J. A., Ivarie, R. D., Morris, J. A., Ullrich, A., Baxter, J. D. and Goodman, H. M. (1978) *Nature (London)* **276**, 795–798.

Sermonti, G. (1959) *Annals N.Y. Acad. Sci.* **81**, 850–973.

Shier, W. T., Rinehart, K. L. and Gottlieb, D. (1969) *Proc. natl. Acad. Sci. USA* **63**, 198–204.

Smith, G. L., Mackett, M. and Moss, B. (1983) *Nature (London)* **302**, 490–495.

Smith, G. R. (1985) Site-specific recombination. In *Genetics of Bacteria*, eds. J. Scaife, D. Leach and A. Galizzi, Academic Press, London.

Spier, R. E. (1987). Processes and products dependent on cultured animal cells. In *Basic Biotechnology*, eds. J. Bu'Lock and B. Kristiansen, Academic Press, London, 509–524.

Sprague, G. F., Blair, L. C. and Thorner, J. (1983). *Ann. Rev. Microbiol.* **37**, 623–630.

Thompson, R. (1986) *J. Antimic. Chemother.* **18**, Suppl. C. 13–23.

Ullrich, A., Shine, J., Chirgwin, J., Pictet, R., Tischer, E., Rutter, W. J. and Goodman, H. M. (1977) *Science* **196**, 1313–1319.

Valenzuela, P., Medina, A., Rutter, W. J., Ammerer, G. and Hall, B. D. (1982) *Nature (London)* **298**, 347–350.

Villa-Komaroff, L., Efstratiadis, A., Broome, S., Lomedico, P., Tizard, R., Naber, S. P., Chick, W. L. and Gilbert, W. (1978) *Proc. natl. Acad. Sci. USA* **75**, 3727–3731.

Windass, J. D., Worsey, M. J., Pioli, E. M., Pioli, D., Barth, P. T., Atherton, K. T., Dart, E. C., Byrom, D., Powell, K. and Senior, P. J. (1980) *Nature (London)* **287**, 396–401.

4 Application of the principles of fermentation engineering to biotechnology

M. A. WINKLER

The development of a successful, large-scale production process is the result of accelerating and intensifying an original concept, usually a small-scale or laboratory process. Operating a large-scale process is not just a matter of carrying out the original procedure with larger quantities of material, but must include new concepts to deal with the problems arising from large-scale operation. As a result, the large-scale version of a process may use techniques and materials considerably different from those used in the original small-scale process. Biological processes present a particular problem in that they operate with very dilute process streams of raw materials and products, even after intensification.

Development activity in biological processes is concentrated in three principal areas: organism development, medium development and fermentation engineering. The development of efficient, high-yielding strains of organisms has been discussed in Chapter 2. In medium development, the formulation of biological growth media is kept continually under review. Roughly three-quarters of the operating costs in fermentation is due to the cost of the growth medium, and alternative sources of biological nutrients are investigated so that advantage can be taken of fluctuations in commodity prices to meet the nutrient requirements of a process as cheaply as possible.

The methods of testing different formulations and finding the 'best' combination of ingredients is described by Winkler (1988). Commercial fermentations use growth media that include materials whose composition is both complex and indeterminate. These may be expensive, specially-developed preparations, such as protein hydrolysates, or by-products of other biological industries. Molasses, a by-product of cane- and beet-sugar refining, provides a source of sugars and other key minor nutrients in the production of citric acid and baker's yeast. Corn-steep liquor (CSL), a by-product of the production of starch and glucose from maize, is used as a source of organic nitrogen and other key minor nutrients, notably in antibiotic production. The composition of both these materials varies seasonally and from one production batch to another. The problems of utilizing waste materials in fermentation processes has been discussed by Winkler (1983).

Fermentation engineering is the branch of biotechnology dealing with the

Table 4.1 Nomenclature

A	Area, m^2
a	Specific surface, $m^2 \cdot m^{-3}$
C_L	Bulk liquid dissolved oxygen concentration, $g\,m^{-3}$
C^*	Equilibrium dissolved oxygen concentration, $g\,m^{-3}$
ΔC	Dissolved oxygen concentration difference, $g\,m^{-3}$
D	Characteristic linear dimension, m
D_i	Impeller diameter, m
DO	Dissolved oxygen
d_B	Average bubble diameter, m
G	Volume gas flow-rate, $m^3\,s^{-1}$
K	Consistency coefficient
K_c	Casson viscosity, $kg^{1/2}\,m^{-1/2}\,s^{-1/2}$
K_s	Saturation coefficient in Monod equation, $kg\,m^{-3}$
k	Constant
k_d	Product decay-rate coefficient, s^{-1}
k_L	Mass-transfer coefficient, $m\,s^{-1}$
$k_L \cdot a$	Volumetric or overall mass-transfer parameter, s^{-1}
k_p	Product formation-rate coefficient, $kg(product)\,kg(biomass)^{-1}\,s^{-1}$
M	Specific growth-rate coefficient, s^{-1}
N	Impeller speed, s^{-1}
N_a	Aeration number, G/ND_i^3
N_p	Power number, $P/\rho N^3 D_i^5$
n	Flow behaviour index
OTR	Oxygen transfer-rate, $g\,m^{-3}\,s^{-1}$
P	Power input, W
P_g	Gassed power input, W
P_p	Process productivity, $kg(product)\,m^{-3}\,s^{-1}$
P_u	Ungassed power input, W
p	Product concentration, $kg\,m^{-3}$
Q	Heat transfer rate, W
q	Heat transfer rate per unit volume, $W\,m^{-3}$
R, R', R''	Empirical coefficients
Re	Reynolds's number, $Du\rho/\eta$
Re^*	'Power-law' Reynolds number (eqn 4.33)
S	Concentration of growth-controlling substrate, $kg\,m^{-3}$
ΔT	Temperature difference, K or °C
t	Time
t_m	Mixing time
t_d	Batch cycle down-time, s
U	Heat transfer coefficient, $W\,m^{-2}\,K^{-1}$
u	Characteristic velocity, $m\,s^{-1}$
V	Volume, m^3
v_s	Superficial gas velocity, $m\,s^{-1}$
w	Biomass concentration, $kg(dry\,weight)\,m^{-3}$
w_m	Maximum or limiting biomass concentration, $kg(dry\,weight)\,m^{-3}$
w_0	Initial biomass concentration, $kg(dry\,weight)\,m^{-3}$
Y	Growth-yield coefficient, $kg(biomass\,dry\,weight)\,kg(growth-controlling\,substrate)^{-1}$
x, y	Empirical exponents
γ	Shear-rate, s^{-1}
$\bar{\gamma}$	Average shear-rate, s^{-1}
ε	Fractional volume of particles
η	Viscosity, $kg\,m^{-1}\,s^{-1}$
η_a	Apparent viscosity, $kg\,m^{-1}\,s^{-1}$
Φ_m	Mixing-time factor

Table 4.1 Nomenclature (continued)

ρ	Density, $kg\, m^{-3}$
τ	Shear stress, $kg\, m^{-1}\, s^{-2}$
τ_0	Yield stress, $kg\, m^{-1}\, s^{-2}$
ζ	Plastic viscosity, $kg\, m^{-1}\, s^{-1}$
μ	Specific growth-rate, s^{-1}
μ_m	Specific growth-rate with unlimited availability of growth-controlling substrate, s^{-1}

design, development, construction and operation of the plant and equipment used in industrial biological processes.

It should be stressed that these three areas of development are not independent, and that the overriding criterion is the final unit cost of the product. For example, a high-yielding strain of organism may have specific nutrient requirements which make the cost of its growth medium outweigh the extra yield of product obtained. Similarly, a cheap raw material may increase the final cost of the product if it is difficult to handle or store, gives a low yield or interferes with product recovery processes. It is worth noting that these strictures apply just as much to growing macroorganisms, in battery chicken, dairy- or fish-farming, as in the cultivation of microorganisms. These are effectively processes for upgrading protein from a cheap, unattractive form into a valuable, acceptable form. For example, although fishmeal is one of the cheapest protein sources for chickens, excessive proportions of this in their diet can produce an off-flavour in their meat. In industrial microbiology, the penicillin production process provides a good example of successful development in all three principal areas, using an organism, cultivation process and nutrient supply all entirely different from those originally used. Most processes, chemical as well as biological, are initiated in the laboratory, involving perhaps a few hundred grams of material, whereas production processes may deal with several thousand tonnes of material. Much of fermentation engineering is thus taken up with identifying and attempting to solve the problems arising from large-scale processing.

These problems can be classified into two broad groups: first, those arising from having to handle large quantities of material, and second, those where the nature of the process itself is affected by the size of the operation—the classic problems of 'scale-up'. In the first group, problems arise because an operation which is simple to carry out in the laboratory may involve complex and expensive equipment where large quantities of material are treated. For example, the small volume of liquid in a typical laboratory flask can be heated by placing it in a water-bath or incubator, but heating several hundred tonnes of liquid involves pumping it through specially-designed heat exchangers. In the second group, 'scale-up' problems arise because the various parameters affecting a process change in different ways as the size or scale of the operation increases. This is discussed in section 4.6.

4.1 The fermenter

The heart of a fermentation process is the fermenter. A working definition of a fermenter is 'a container in which is maintained an environment favourable to the operation of a desired biological process'. The word 'fermenter' tends to conjure up visions of large, shiny, stainless-steel vessels bristling with sophisticated instrumentation. While many industrial fermenters are indeed like that, several important large-scale biological processes are carried out in equipment which is considerably less sophisticated—and so considerably cheaper. In fermentation engineering, it is always useful to bear in mind the principle of Occam's Razor, inferred from the writings of the 14th-century Surrey philosopher, William of Ockham: *entia non sunt multiplicanda praeter necessitatem*. In the context of fermentation engineering, this can be paraphrased as 'systems should be no more complicated than really necessary'. For some processes, a fermenter consisting of a simple open pit or tank is quite satisfactory, whereas for processes very sensitive to the conditions in the fermenter, then a closed, closely-controlled system is needed to maintain a favourable environment.

4.1.1 *The environment*

The environment maintained in a fermenter can be considered under three headings: biological, chemical and physical.

The biological environment is favourable to a biological process when only desirable organisms, contributing to the process, are present, and undesirable, destructive, non-productive or inefficiently productive organisms are excluded. From the production point of view, it also implies that the desired organisms are present in sufficient quantity to carry out the desired process at an economical rate, although this is usually assumed to result automatically from the maintenance of a favourable overall environment. A system containing only a selected desired organism is said to be 'aseptic'. Asepsis is achieved in practice by rendering the system free of all living organisms, or 'sterile', and then introducing the desired organism. Introduction of an organism into a growth system is called 'inoculation', and the distinction between 'aseptic' and 'sterile' should be noted. As will be seen in sections 4.2.1, 4.3.2 and 4.3.3, aseptic conditions are expensive to attain and maintain in terms of plant design, construction and operation, and should be no more stringent than the process really requires. This may involve the fermentation engineer in risk assessment and balancing the risk of contamination of a process, the cost of contamination and the cost of preventing contamination.

The chemical environment involves the composition of the microbial growth medium, which should contain the desired concentration of substrate or microbiological nutrients and synthetic precursors, free from inhibitory substances and maintained at the desired pH. While most soluble nutrients are supplied in the medium make-up stage before charging to the fermenter, maintaining an appropriate level of dissolved oxygen is one of the major problems confronting the fermentation engineer. The low solubility of oxygen gas means that it must usually be supplied continuously even in a batch process. This is discussed in detail in section 4.5. For anaerobic and anoxic processes, oxygen is excluded as it is inhibitory to the process, and in some cases may give rise to an explosion hazard where a combustible substance, such as methane, is produced by the process. Process products may be inhibitory to the process itself, notably carbon dioxide, and it may be necessary to arrange their continual removal. Other parameters of fermentation of possible relevance are the low water-activity in the medium, due to the concentration of solutes, and ionic strength, due to ionic substances such as dissolved salts.

The physical environment refers principally to the temperature of the system, and temperature control is an important consideration in fermenter design. This and the desirability of maintaining uniform conditions throughout the fermenter involve mixing, which in turn creates mechanical shear that can disrupt organisms and their agglomerates, such as flocs or pellets. It should be noted that in maintaining a favourable environment, environmental parameters are not necessarily held constant with time. Batch fermentations pass through different phases of growth which may have different environmental optima, so that a fermentation may follow a 'profile' or programme of changing nutrient availability, pH or temperature, to favour successive growth phases.

4.1.2 *Principal types of fermenter*

Several different types of vessel are used for large-scale biological processes, and their degree of sophistication in design, construction and operation is determined by the sensitivity of the process to the environment maintained in the vessel. In some long-established processes, such as brewing and penicillin manufacture, fermenter design has evolved from that of a simple container into more complex modifications as understanding of the biological process mechanisms has been extended. In the simplest case, the fermenter is a vessel in which reagents, substrates and organisms are brought into contact with provision for their addition and removal. Fermenters can then be classed into two broad groups, *suspended-growth* systems and *supported-growth* systems, although most fermenters contain elements of both systems. In suspended-growth systems, the organisms are immersed in and dispersed throughout their nutrient medium and their movement follows that of the nutrient liquid. In supported-growth systems, the organisms grow as a layer or film on a surface in contact with a nutrient medium. In practice however, suspended-growth systems have a film of organisms on the surfaces of the container, and supported-growth systems usually have organisms dispersed in the nutrient medium.

4.1.2.1 *Suspended-growth systems.* The simplest type of fermenter is an open tank in which the organisms are dispersed into nutrient liquid. These have been used successfully in the brewing industry for generations, and in the anaerobic stage of the fermentation, a foam blanket of carbon dioxide and yeast develops which effectively prevents access of air to the process. Cooling coils can be fitted for control of temperature during fermentation. Open concrete pools can be used, for example in the preliminary stage of the production of synthetic hormones, where chopped yams are fermented to release a steroid from its glycoside by enzymatic hydrolysis. Open pools or 'lagoons' are widely used for low-rate biological waste-water treatment, where the liquid surface allows dissolution of oxygen from the air and escape of carbon dioxide. These effects can be accelerated by agitating the liquid, which enhances the transfer of gases, and the maintenance of homogeneity. This can be carried out with a mechanical stirrer or by the rise of gas-bubbles through the liquid. With aerobic processes, the sparging of air into the liquid provides oxygen for the process as well as agitation. In anaerobic processes, gases released by the fermentation can provide agitation, using a pump to recirculate the gas through a sparger, or by arranging for the gas bubbles evolved during fermentation to create agitation during their rise through the liquid. The latter effect is utilized in the 'conical' fermenter used in brewing, which is a tall cylinder with a conical base section (Figure 4.1). For aseptic conditions, closed tanks must be used, and conventional fermenters for aerobic processes are closed vessels, sparged with air, which usually have additional mechanical stirrers.

4.1.2.2 *Supported-growth systems* (surface-culture films and fluidized beds). Growing microorganisms on the surface layer of nutrient medium held in a dish or tray is a standard laboratory technique, called 'surface culture'. Penicillin was originally produced by surface culture in thousands of bottles kept in an incubator, before being superseded by deep-tank culture. Surface culture is still used for industrial production of citric acid by the mould *Aspergillus niger* and for itaconic acid by *A. terreus* (Kristiansen and Bu'Lock, 1980). The mould is grown on the surface of a suitable medium such as molasses in shallow trays, kept in stacks in a constant-temperature room with air blown over the trays to provide oxygen. This method is useful in utilizing low-grade molasses.

Microbial films can be developed on the surfaces of a suitable packing medium. This can be in the form of a fixed bed, of stones, plastic pieces or ribbed plastic sheets, through which nutrient liquid is trickled to contact the microbial film on the packing surfaces. This system is widely used in

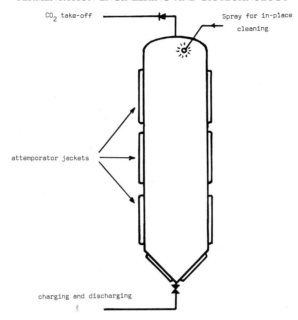

Figure 4.1 'Conical' brewing fermenter.

biological waste-water treatment, and for aerobic processes, the liquid flow-rate is normally kept low enough to leave room in the packing for circulation of air to supply oxygen. For anaerobic or anoxic processes, the packing is flooded and held in a closed container. Such closed systems can also be used for aerobic processes, with air sparged directly into the flooded packing or by pre-dissolving high-purity oxygen into the nutrient liquid feed. Vinegar can be made by trickling dilute ethanol solution, beer or wine, through a bed of beechwood shavings on which a culture of *Acetobacter* develops. As an alternative to trickling the nutrient liquid through the packing, the packing can be moved through the liquid. This moving bed approach can be effected in two ways: as a fluidized bed or as a coherent moving bed.

In *fluidized beds*, the biological film is developed on particles which are suspended in an upward flow of liquid in which they are then free to circulate. A fluidized bed thus has features of both suspended- and supported-growth systems. The support particles can be solid, such as sand or glass beads, or porous, such as plastic or stainless-steel mesh. Porous particles allow growth within the particle as well as on the surface, and they have been used successfully in pilot-scale production of citric acid by *Aspergillus foetidus* (Atkinson and Lewis, 1980). The advantages of fluidized beds is that they avoid blockage of the packing by luxuriant biological growth, which happens with fixed packed beds, and the loss of biomass by 'wash-out' which can happen in continuous suspended-growth systems. Where organisms can be induced to form large aggregates or flocs, a fluidized bed can be developed without a support medium. This has been used with some success in the tower fermenter, using a flocculent strain of yeast, and also in waste-water treatment, but with the risk of the organisms reverting to a non-flocculating form and being carried out of the system with the output stream. Oxygen supply to fluidized beds can present problems, and they are most readily used in anaerobic processes, such as brewing, or in anoxic processes, such as denitrification, or where the oxygen demand is very low, as in drinking-water treatment. For aerobic systems, oxygen is most conveniently supplied by pre-dissolution of high-purity oxygen in the feed-stream to the process. Fixed- and fluidized-bed systems are also used in immobilized-enzyme and immobilized-cell reactors (Chapter 7). *Coherent moving-bed systems* have the packing with its associated microbial film mounted on a shaft and rotating in a bath of nutrient medium. The packing may be a fixed-bed type held in a cage or in the form of discs. Closed versions of this have been developed for aseptic aerobic operation (Anderson and Blain, 1980) and for anoxic denitrification of waste-water (Winkler, 1981).

4.1.3 *Mode of operation*

Reaction systems can be run as continuous systems or as batch systems, and not all of the types of fermenter discussed in the previous section are suited to both batch and continuous modes.

In *batch operation* (see 2.2.2), the reactor is charged with the reacting species, and as the reaction proceeds, the conditions in the reactor change as reagents are consumed and products are formed. When the desired degree of reaction has taken place, the reactor is discharged and cleaned and the process repeated. Batch systems are thus non-steady-state processes.

In *continuous operation* (see 2.4.6.1), fresh reagents flow continuously into the reactor and the product stream flows out continuously. The events in the reactor itself can follow either of two basic patterns, (i) plug-flow or (ii) completely-mixed. In a plug-flow continuous system, the nutrient medium is inoculated with microbial culture on entry to the reactor, and the organisms carry out their biological activity as the liquid flows through the system, and pass out of the reactor with the spent medium. The organisms may then be separated from the product stream and recycled to inoculate the inlet feed stream. In the form of open channels, this system is widely used in waste-water treatment, and closed tubular fermenters are being developed for antibiotic production. Tubular enzyme reactors are used for the saccharification of starch to sugars, such as in continuous mashing of malt and commercial production of glucose from starch. In a plug-flow system, each element behaves as a miniature batch reactor as it passes through the system, and is thus a non-steady-state operation. With a uniform liquid velocity, the distance along the length of a plug-flow reactor corresponds to the time-scale of a batch reactor, and conditions at any particular point remain constant with time.

In a completely-mixed continuous system, conditions are ideally uniform throughout the reactor in an equilibrium mixture of nutrients, organisms and products. The feed to the system is organism-free nutrient and in some cases an inoculum of recycled organisms. The outlet stream is the equilibrium mixture from the fermenter and emerges at a rate such that the loss of organisms in the outflow balances those added to and grown in the fermenter. If the loss-rate exceeds the growth-rate, the condition is called 'wash-out' and results in the complete removal of organisms from the fermenter. Organisms are in some cases separated from the outflow stream and recycled to the inlet to maintain a high microbial concentration in the reactor. In a completely-mixed continuous system, conditions at equilibrium average out to the conditions in a batch reactor 'frozen' at a particular instant in time. This point on the time-scale depends on the reactor configuration and the liquid flow-rates, but for a simple straight-through system with no recycle of organisms, for example, it corresponds to the mean hydraulic residence time in the system. This is an average over a statistical distribution pattern of residence times for individual liquid elements, and some elements will pass directly from the inlet to the outlet, carrying unconsumed nutrients with them. Other elements will remain in the reactor long after their reagents have been consumed. To save the time needed for the system to achieve equilibrium, completely-mixed continuous systems are usually started up as batch systems and then switched over to continuous operation when conditions are close to the required holding-point.

In practice, no system is perfectly plug-flow or perfectly completely-mixed. In a plug-flow reactor, successive liquid elements mix due to turbulence and solute diffusion, and perfect mixing of biological systems is difficult due to the presence of several different phases—nutrient liquid, organisms, gas bubbles and often solid nutrient particles as well. Some designs of fermenter combine features of plug-flow and completely-mixed systems, such as the air-lift fermenter and continuous-channel waste-water treatment systems.

'Serial' or 'fed-batch' operation can be considered as combining batch and completely-mixed continuous fermentation, as a batch process is prolonged by feeding in additional nutrient medium continually or intermittently during the course of the fermentation. The object of this procedure is to maintain the specific growth rate of the organism at the level giving maximum productivity, by controlling the nutrient concentration. In penicillin production, intermittent feeding is carried out in response to the dissolved-oxygen (DO) level in the growth medium. This falls as the biomass concentration increases, as a result of the concomitant increase in oxygen uptake rate, and also because the biomass growth increases the viscosity of the medium, which reduces the oxygen transfer rate. Some culture is withdrawn, and fresh medium is injected. In baker's yeast production, the nutrient medium is fed in continuously, either according to a pre-calculated programme, or with the injection rate adjusted in response to estimates of nutrient consumption computed from exhaust-gas analysis (Winkler, 1988).

In comparing the relative benefits and problems of batch and continuous processing, the principal disadvantages of batch processing can be summarized as the high proportion of unproductive 'down-time' in fermenter operation, difficulty in the design and operation of non-steady-state processes, batch-to-batch variability, uneven demand on services and utilities and the accumulation in the process of inhibitory products. The principal advantages are a lower contamination risk, operational flexibility where fermenters are used for the manufacture of several different products, the ability to run different successive phases in the same vessel, closer control of the genetic stability of the organism, the ability to identify all materials involved in making a particular batch of product—vital in the pharmaceutical industry—easier coordination with previous and subsequent batch processing stages and the ability to hold an organism in a declining growth-phase. Thus continuous fermentation is best suited to a growth-associated process with a low risk of contamination and a genetically stable organism, giving a product for which the demand or storage characteristics justify long production runs, or whose accumulation in the process is inhibitory.

Fermenters must be emptied, cleaned, sterilized and recharged before each fermentation, all essential but non-productive operations. In batch processing, these can take up nearly as much time as the fermentation itself; in batch penicillin production, for example, each batch takes a total of about 180 h, of which the fermentation takes only about 100 h. In continuous processing, however, a run can last several weeks or even months, so that the proportion of non-productive time is small. The continual inflow and outflow of nutrient medium in continuous processes do provide two additional risk points for contamination, although the continual removal of products in the outflow does prevent the accumulation of inhibitory products in the system. Completely-mixed continuous systems are steady-state systems and require relatively unsophisticated control equipment, but they depend on a balance being maintained between addition, growth and removal of organisms. This can be difficult to stabilize when the organism is most productive in a declining growth phase, as is the case with penicillin production. In addition, the advent of microprocessors has considerably simplified the design and control of non-steady-state systems, and control programmes can be developed to adjust process conditions to the environmental optima for different successive growth phases through which many biological processes pass.

This is called 'environmental time-profiling', and may be effected by pre-calculating optimal trajectories for each of the process control variables, by programming a control system to recognize the advent of a new growth phase and adjust the conditions to a new optimal set for that phase, or by continually estimating the state of the process, using a microprocessor, and adjusting the control variables to maintain a desirable calculated value of one of the organism growth parameters, such as specific growth rate or respiration quotient. This is called 'adaptive control'. With a completely mixed system, different environments must be contained in separate reactor volumes. Some ancillary processes are inevitably batch operations, notably the make-up of nutrient medium, so that a freshly-prepared batch of medium can be charged to a batch system with minimal delay. The interface between batch and continuous processes has to be a buffer or holding vessel in which long residence times involve a risk of deterioration. In addition, in pharmaceutical manufacture it is essential to be able to identify the origin of all material involved in the production of any particular unit of product, which is practicable only with batch operation. In general, continuous operation is not widely used in the biological industries, except for waste-water treatment, where there are no contamination problems, and production of single-cell protein, which is essentially a growth-associated product. In this context it is interesting to consider why baker's yeast is produced in batch or serial processes, as, at first sight, it is biomass production with a low infection risk ideally suited to completely-mixed continuous fermentation. In practice, the strain of yeast is important for its bread-making ('fermentation') characteristics, and the yield of yeast is strongly affected by the pattern of incremental feeding (Burrows, 1979).

4.1.3.1 *Unit size.* The total capacity of a fermentation plant is determined by the expected demand for its products—often little more than an informed guess—and the productivity of the process. The engineer then has to decide whether this capacity should be installed as a large number of small units or a small number of large units. In some cases, the unit size may be influenced by non-technical factors. For example, fabrication of a vessel in the constructors' workshops gives a higher standard of workmanship than on-site construction, but the size of vessel is then limited to the maximum that can be transported by road or rail. The height of vessel used may then be limited by the permissible ground loading or cost of foundations, or the space available. Apart from this, in general, large units give a lower unit cost of product than small units, particularly with highly instrumented systems, as the cost of instrumentation is much the same for small vessels as for large ones. Small vessels are favoured where a wide product range is required and where the risk of breakdown is serious—for example, with only one vessel, one breakdown represents 100% loss of production.

4.1.3.2 *Productivity.* The total production capacity required in a fermentation plant is worked out by comparing the required production rate with the productivity of the process. The productivity of the fermentation stage is the amount of product produced per unit time per unit volume, and in a batch system is the product concentration divided by the time taken to achieve that concentration. The total time must include the batch preparation or 'down' time, for discharging, cleaning, re-charging and sterilizing, as well as the actual fermentation time. In a batch system, the product concentration varies continuously, and thus so does the productivity. The point of maximum productivity, where the fermentation is stopped and the next batch cycle started, can be selected from plots of the product concentration against time, which can usually be conveniently represented by a mathematical model or correlation.

Biomass growth in many systems can be described by the 'logistic' equation:

$$\frac{dw}{dt} = M \cdot w \left(1 - \frac{w}{w_m}\right) = \frac{M \cdot w}{w_m}(w_m - w) \qquad (4.1)$$

where w is the biomass concentration, t is time, M is a specific growth-rate coefficient and w_m is the maximum or limiting value of w obtainable in the system. The specific growth-rate of the organism starts with a value close to M and decreases thereafter as the biomass concentration approaches its limiting value w_m asymptotically. It is a very useful model, as it makes no assumptions about the nature of the substrate on which the organism is growing, and is used successfully to describe organisms growing in complex media. The logistic equation can be integrated to give an expression explicit in the biomass concentration:

$$w = \frac{w_0 w_m e^{Mt}}{w_m + w_0(e^{Mt} - 1)} \qquad (4.2)$$

where w_0 is the biomass concentration when $t = 0$. The values of M, w_m and w_0 for a given system can be determined empirically, finding the combination of values that gives the curve best fitting the experimental biomass–time data.

In a more closely-specified system, where a single organism is growing in a medium where growth is controlled by the availability of one substrate, and all the other nutrients are in ample supply, then the Monod saturation equation can be applied:

$$\mu = \frac{1}{w}\frac{dw}{dt} = \mu_m \frac{S}{K_s + S} \qquad (4.3)$$

where μ is the specific growth-rate of the organism, S the concentration of the growth-controlling substrate, K_s a saturation coefficient for that organism and that substrate, and μ_m the maximum or asymptotic value of μ achieved when the growth-controlling substrate has unlimited availability. This equation relates the growth-rate of the organism to the availability of one of its nutrients. The production of biomass and the concomitant consumption of the growth-controlling substrate are related by a growth-yield coefficient, Y, defined by

$$Y = -\frac{dw}{dS} = \frac{(dw/dt)}{-(dS/dt)} \qquad (4.4)$$

It should be noted that Y is not dimensionless, but has units of 'quantity of biomass per unit quantity of substrate'. In differential form, the definition is always true, but difficult to use, so that it is usually necessary to assume that Y is constant over the period considered. The definition can then be integrated to give

$$w = w_0 + Y(S_0 - S) \qquad (4.5)$$

or

$$S = S_0 - \frac{(w - w_0)}{Y} \qquad (4.6)$$

where w_0 and S_0 are the values of w and S when $t = 0$. It also follows that the maximum biomass concentration obtainable in the system, w_m, is given by

$$w_m = w_0 + Y \cdot S_0 \qquad (4.7)$$

The equations (4.4), (4.6) and (4.7) can be combined with the Monod equation (4.3) and integrated to give a relation between biomass concentration and time:

$$\mu_m \cdot t = \ln\frac{w}{w_0} + \frac{Y \cdot K_s}{w_m} \ln\left(\frac{w(w_m - w_0)}{w_0(w_m - w)}\right) \qquad (4.8)$$

This relation has the advantage that the key growth kinetics coefficients, μ_m, K_s and Y, are all characteristic of the organism in the prevailing conditions of pH, temperature and so on, and with that growth-controlling substrate, and can, in principle, be determined in, and valid for, a separate system. They can also, of course, be found by curve-fitting with experimental data, as with the logistic equation.

The formation of secondary metabolites, such as antibiotics, has been successfully modelled by taking the rate of their formation as proportional to the biomass concentration, and the rate of their degradation as proportional to their concentration, so that

$$\frac{dp}{dt} = k_p \cdot w - k_d \cdot p \qquad (4.9)$$

where p is the product concentration, and k_p and k_d the product formation- and decay-rate coefficients (Winkler, 1988). By using a suitable expression for the biomass concentration, such as eqns (4.2) or (4.8), the product concentration at different times throughout the batch can be computed by numerical integration. Secondary metabolites, such as antibiotics, are not generally growth-associated, so that the product concentration starts to rise a considerable time after the start of the fermentation. A product 'lag' can then be included in the computation. This can be a fixed time-lag, from practical experience, or one determined by some other process phenomenon, such as the specific growth rate of the organism falling to a pre-set proportion of its maximum value. The productivity of the process, P_p, is then given by

$$P_p = \frac{p}{t + t_d} \quad (4.10)$$

where t_d is the 'down-time' in the batch cycle, and p is the product concentration at a time t after the start of fermentation. The productivity is thus the slope of a line from the start of the batch cycle to the point on the product curve. Most batch product curves tend to level off towards the end of the batch, so that the point of maximum productivity is where the line from the start of the batch makes a tangent with the product curve (Figure 4.2).

This also demonstrates that, if the down-time is increased, then the fermentation time also needs to be increased to maintain maximum productivity for the batch. An increase in down-time could result from a pump breakdown causing an increase in charging and discharging times, or a steam supply fault necessitating an increase in sterilization time, for example. The actual length of batch times may also be modified to fit into a convenient operational period, such as a working week.

4.1.4 *Ancillary processes*

Although the fermenter is the heart of a biological production process, its successful operation is dependent on a number of ancillary units. The range varies with each process, but all microbiological processes require make-up, sterilization and cooling of nutrient medium and inoculum preparation. Aerobic processes require air compression and sterilization and safe dispersion of exhaust air. On completion of fermentation, the culture broth is discharged and passed to a series of product recovery operations, the first of which is usually separation of the biomass from the spent medium, by filtration, flocculation or centrifuging. Extracellular products are separated from the broth by, for example, solvent extraction, fractional precipitation or distillation. Intracellular products within the microbial cells must be released by cell smashing or lysis before product recovery. The fermenter itself requires steam, cooling water and environmental control fluids for pH and foam

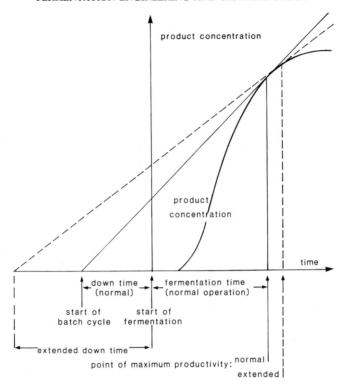

Figure 4.2 Maximum batch productivity.

control, which require sterilization. The seed fermenter used for inoculum preparation is a small-scale version of the production fermenter, requires the same services, and may itself be inoculated from a smaller seed vessel. Volumes of inoculum used are typically about 10% of the volume of the culture inoculated.

4.2 General design principles

Environment-sensitive processes require a closed and controlled environment to be maintained within the fermenter, but with aseptic entry and outlet for nutrient media, air and control fluids and provision for heating, cooling and agitation. While the requirements for exclusion of contamination can be extremely demanding, fermentation conditions are undemanding in terms of temperature, pressure and reagent corrosion compared to chemical reactors. Considerable experience has now been accumulated on a small number of designs of fermenter, and the simplicity of fermenters gives them a versatility enabling virtually any fermenter to be used for a range of biological processes provided it meets the required aseptic standard.

The design of fermenter selected for a particular process is often not necessarily optimal, but chosen for reliability and predictability of performance over an untried design of possible (but unproven) greater efficiency. This also has the advantage that in plants where different-sized units of similar design are used, experience gained on one unit can frequently be utilized on another. However, the current intensive development activity in biotechnology will doubtless result in specialized designs for novel processes, whose technology will be used in modifying more conservative designs.

4.2.1 Basic design rules

The basic rules for the design and construction of aseptically operated plant are based on the precept that microorganisms are extremely small, with dimensions much less than normal engineering tolerances. This means, for example, that what would be regarded as a satisfactory tight closure in a non-biological system could afford ample space for the passage of microorganisms. Strict application of the rules for aseptic operation does, however, incur expense and inconvenience in construction, operation and maintenance of the plant, and should be applied only after a realistic assessment of the risk and resultant cost of contamination. For example, stringent precautions to preserve strictly aseptic conditions have been justified in the continuous production of bacterial biomass from methanol for animal feed (Andrew, 1981), whereas completely aseptic operation is not considered necessary in serial batch production of baker's yeast (Burrows, 1979).

For strictly aseptic operation, the mechanical integrity of the plant must be maintained. Entry points to the aseptic region of the plant, such as mechanical seals for agitator and pump shafts, valve closures, probe insertions, sample ports and joints, are contamination risks and should be as few in number as possible. The structure should as far as possible be all-welded, with no mechanical closures such as flanged or screw joints that can be replaced by welded joints. The resulting maintenance problems, where removal and replacement of a component involves cutting and re-welding, is justified by the avoidance of contamination (Smith, 1980) as mechanical joints work loose due to vibration and the thermal expansion and contraction in sterilization cycles. Where such closures are unavoidable, they should be surrounded by a steam box. All connections to the aseptic region of the plant should be steam-sealed, including the glands of the valves used to control liquid flow in the pipe and steam flow to the steam lock (Andrew, 1981), and there should be no direct connection between sterile and non-sterile sections of the plant. Simple examples of steam locks are shown in Figure 4.3.

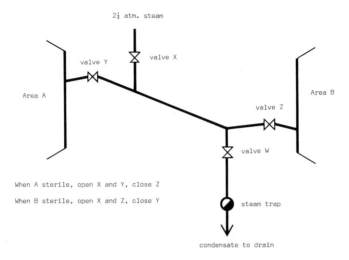

Figure 4.3 Simple example of steam-lock.

Pipelines should be kept under steam when not in use and should be constructed to slope to a definite drain point so that stagnant liquid cannot accumulate at any point. Steam is fed in at the top of the slope and the condensate run out through a steam-trap at the drain point. In some operations, the drain is kept open so that steam can be seen to emit from the drain-point, but this wasted steam creates an unpleasant working atmosphere. Recent trials in a penicillin plant showed that steam traps save steam and also give a lower infection rate, but they do need regular checking. The general design of the plant should also avoid stagnant regions, dead spaces, pockets, pipe branches and crevices which can not only collect stagnant liquid and microorganisms, but are also difficult to sterilize effectively. Such areas are also more liable to corrosion.

4.2.2 *Materials and components*

The standard construction material for production-scale fermenters is a suitable grade of stainless steel. Different formulations have different mechanical properties, corrosion resistance, ease of fabrication, availability and cost. Stainless steel is about ten times more expensive than ordinary mild steel., and is more difficult to machine and weld, but is stronger, so that stainless-steel vessels can be self-supporting where a similar mild steel vessel would need a separate support structure. The corrosion resistance of stainless steel is due to a tough surface film of chromium oxide. This is resistant to oxidizing conditions, but is liable to damage by reducing agents, chloride and abrasion. Resistance to chloride and sulphur dioxide can be increased by the inclusion of about 3% molybdenum in the alloy, and the film is stabilized during welding by inclusion of titanium or niobium in the alloy. Stringent washing may damage the film, which can be replaced by a wash with an oxidizing acid such as nitric acid, which in turn is rinsed out with clean water. Particulate materials in the growth medium can also abrade the oxide film during processing. To ensure coherent oxide film formation, the alloy should contain about 18% chromium, and corrosion resistance is enhanced by stabilizing the non-magnetic austenitic structure in the steel by the inclusion of 8% nickel, so that such alloys are referred to as '18/8' stainless steels. A cheaper alternative to stainless steel is mild steel lined with a protective coating of glass, rubber, synthetic resin or thin stainless-steel sheet. However, difficulties in bonding between the lining and the steel can arise and the lining may be damaged during plant modification or installation. For non-aseptic operation, cheaper and more robust materials can be used, such as concrete, brick or glass-reinforced plastic.

Valves are needed to regulate fluid flow, and should be capable of handling hot and cold liquids, possibly containing suspended solids, and steam. They represent an infection risk point, since they inevitably involve a joint or mechanical closure. Diaphragm valves are widely used in fermentation, or, for strictly aseptic conditions, a piston valve with a steam-sealed gland.

Pumps should as far as possible be avoided (fluid movement can be effected by gravity and/or sterile gas pressure) as they involve moving components which can provide a contamination risk point, although the risk can in some cases be reduced by interposing a sterilizing stage between the pump and the sterile area. Centrifugal pumps are commonly used in biological industries, the seal on the rotating drive shaft being the risk point, and for strictly aseptic operation a steam-sealed piston pump can be used.

For flexibility of operation, the main sections of plant should be capable of independent sterilization, but must be connected through a steam lock to prevent unsterilized plant being connected to sterile plant. In the simple steam lock shown in Figure 4.3, the valves themselves are protected by steam passed through their glands, and the pipeline connecting units A and B is kept sterile by steam passing through valves X and W, the condensate passing out through the steam trap. For sterilizing A, the valve Z to unit B is closed and valve X to A opened. When A is sterile, valve Y is closed, and Z opened to admit steam to B. For transfer of fluid between A and B, X and W are closed and Y and Z opened. For sterile operation, this sequence must be followed precisely, and even this simple example entails half-a-dozen separate operations. A full-scale plant may have a large number of such locks, so that sterilization of units and transfer of material round the plant involves hundreds of sequential operations of valves, instruments and pumps. In such a case, computer control is virtually essential to ensure that the precise sequence is observed and that no valve may be operated before correct completion of the preceding sequence. Such a control system has been installed in the ICI continuous SCP plant (see 2.4.6.10), where failure to follow precise operational sequences was found to lead to incomplete sterilization and process failure due to contamination within a few days (Smith, 1980). It is arranged so that the sequence can be altered only at a very high level and cannot be overridden by, for example, operating a valve manually. The strictly aseptic conditions are also continually monitored in a 'sterile audit' of the physical state of the plant, operational sequences, operator training, plant records and engineering integrity of the plant. This involves checks on the sterile closures, leaks at closures, valves and welds, steam-straps, temperature records, sterility risk reports, records and actions, and maintenance and line modifications (Smith, 1980).

4.2.2.1 *Air sterilization.* The sterilization of air for oxygen supply to an aerobic process is not only vital but also extremely demanding. A typical batch fermentation could be supplied with several hundred tonnes of air, which, in principle, should be completely free of microorganisms. Although some thermal sterilization of air may occur during compression, the most usual method is deep filtration, where the air is passed through a pad of fibrous material. The effect of the depth of the pad on particle removal is logarithmic, in that each unit depth of pad removes the same *proportion* of the particles passing into it. This means that although deeper pads remove more particles, in theory an infinite depth is needed to effect complete removal of particles from the air. The pad is routinely sterilized by passing steam through it or by heating it electrically. 'Absolute' air sterilization can be obtained by passing the air through filter media with apertures or pores small enough to prevent the passage of particles of any selected size, down to fractions of a micrometre to trap microbial spores or virus particles. Such filter media, however, are expensive, 'blind' rapidly, require frequent replacement, and give rise to a high gas pressure-drop. Materials used for absolute filter media include paper, unglazed ceramic, resin-bonded paper and porous plastic septa.

4.2.3 *Control and instrumentation*
Several of the environmental parameters in the fermenter, discussed in section 4.1.1, can be monitored by means of analytical probes, the signals from which, after amplification, can then be used to actuate control procedures. Thus, signals from thermistors, monitoring temperature, can be used to regulate the flow of cooling water during fermentation or of steam during sterilization. Similarly, control of pH can be effected by the injection of sterile alkali or acid, under sterile gas pressure, in response to a pH electrode. Where aerobic organisms can utilize ammonia as their nitrogen source, ammonia can be

injected into the air stream prior to sterilization to control pH, so that a nutrient is used as an environmental control. This principle can also be used in the control of the dissolved oxygen (DO) level in a fermentation. Low DO levels can inhibit aerobic cultures, and high DO levels inhibit oxygen dissolution and so waste energy. The signal from a DO probe assembly can be used to regulate the air-flow into the system, the agitation intensity or the flow of nutrient medium into a continuous process. It should be emphasized that an analytical probe responds to the conditions at a particular point, and may not indicate the overall fermenter environment.

Control systems of different levels of sophistication are available and should be selected according to the sensitivity of the process to each parameter. One of the simplest is where regulatory action is initiated when the value of a parameter falls outside the range between two preset values. The correcting action may then be the injection of sufficient control fluid to return the parameter to within the desired range, or injection of a preset quantity of control fluid. More detailed control of fermentations is obtained using microprocessors, which can infer trends from key parameters, monitored automatically at regular intervals, and then initiate corrective action before excessive deviation from the desired values has occurred. Control systems need careful 'tuning' so that the system does not 'hunt' or try to correct every random variation. In an extreme case, hunting can lead to wild cycling of the environment due to successive over-corrections in opposite directions. As biological reactions are comparatively slow, action can be taken on the basis of several data reports, thus reducing the sensitivity to purely random effects. Microprocessors are particularly useful in the control of non-steady-state systems, where the phase of a reaction can be inferred from environmental trends, for changing the control settings appropriate to the different phases of a process. A parameter with a close stoichiometric relationship to the reaction is needed for this. For example, the oxygen and/or carbon dioxide content of the exhaust gas stream, related to the air input of an aerobic process, could indicate the rate of carbohydrate metabolism, and, from a cumulative analysis over the course of the process, the consumption and probable residual concentration of carbohydrate computed. Where no reliable stoichiometric relation is found, the output of automated analysis of small samples, by ion-selective electrodes or spectrophotometry, can be fed to a microprocessor (Winkler, 1988).

Designs of most types of analytical probe are now available which are capable of withstanding steam sterilization and so can be sterilized *in situ* with the fermenter vessel. As the boundary between the inside and outside of the probe may be only a delicate glass or plastic membrane, the pressure due to the steam in sterilization must be balanced by enclosing the probe in a housing pressurized with sterile gas.

A positive pressure is maintained in aseptic fermenters, to reduce the risk of contamination, by means of a control valve on the exhaust gas outlet actuated by a pressure-sensor in the fermenter. The exhaust gas passes through a filter before discharge to prevent microorganisms being carried out into the atmosphere, but a temporary fall in fermenter pressure could draw air into the fermenter carrying microorganisms with it from the exhaust filter, giving rise to 'back infection'. The exhaust air creates a further problem, as it is saturated with water vapour which tends to condense in the gas exhaust and run back into the fermenter. This can be prevented with a drained catch-pot, or by heating the exhaust gas with steam coils before passing to the exhaust gas filter.

Foam control devices are installed in fermenters, because many nutrient media contain materials such as proteins which stabilize the foam generated by gas sparged or evolved during fermentation. Foam is undesirable, as it occupies the fermenter head-space, which needs to be large enough to allow droplets of liquid growth medium to disentrain efficiently from the exhaust gases as they leave the vessel, particularly in air-lift and bubble-column fermenters. Foam can also be carried out with the exhaust gases, resulting in loss of culture and clogging of the exhaust filter. Some materials and organisms can be selectively carried into the foam phase (Thomas and Winkler, 1977) thus removing them from the main body of the culture. Foam can be controlled with mechanical foam-breaking devices or by the injection of foam-suppressing or foam-breaking chemicals. Several patented mechanical foam breaking devices are available and commonly used in smaller fermenters, but are reported as having about the same power consumption as mechanical agitation (Schügerl *et al.*, 1978). Anti-foam fluids can be injected, under sterile gas pressure or by a dosing pump, either at pre-set intervals based on operational experience or in response to a foam detector. In the latter case, a signal is generated when foam reaches a preset level in the fermenter, either by completing the circuit between two electrodes or changing the electrical capacitance between metal tapes. Higher alcohols, such as octanol, act as foam breakers, causing disruption of the liquid film in the foam by reducing surface tension and creating a high surface pressure. Silicones, natural oils and other high molecular-weight esters inhibit foaming by adsorption at the gas–liquid interface, forming a layer, which can however interfere with oxygen dissolution (Schügerl *et al.*, 1978), and impaired yield in some fermentations has been attributed to excessive antifoam dosing (Finn, 1967). Examples of foam inhibitors are polymethylsiloxane, tributyl phosphate, polyethylene glycol, polyoxyethylene–propylene copolymer and lard oil. The synthetic foam inhibitors are expensive but effective in very small quantities such as 0.1%. Lard or vegetable oil used at about 0.5% level in penicillin production acts as both antifoam and nutrient.

Control fluids should be sterile, and the acid and alkali for pH control can be inherently sterile if sufficiently concentrated, but require very efficient and rapid dispersion into the culture. Some complex media are self-buffering, and crude pH control is given by including chalk in the growth medium.

4.2.4 *Stirred-tank fermenters*

Stirred-tank fermenters (Figure 4.4) are agitated mechanically to maintain homogeneity, to attain rapid dispersion and mixing of injected materials, and to enhance heat-transfer in temperature control and mass-transfer in dissolving sparingly soluble gases such as oxygen. The extent to which these aims are achieved depends principally on the power dissipated into the medium by the agitator, so that the agitator is essentially a power transmission device. The effectiveness of the power input then depends on the configuration of the agitator and other fermenter components. The stirred-tank fermenter is a versatile design and is used in a range of sizes from one-litre laboratory units to production-scale vessels of typically 100-tonne capacity, so that considerable operational experience has accumulated with them. The configuration of the fermenter vessel itself is derived from that of

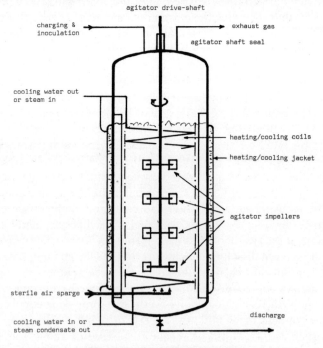

Figure 4.4 Stirred-tank fermenter.

conventional pressure vessels, which are widely used in industrial processes and for which detailed and comprehensive design procedures have been developed. The fermenter is usually constructed as an upright cylinder, with dished ends to facilitate liquid drainage during medium discharge at the bottom, and of splashes and disrupted foam during operation at the top. Dished ends are usually torispherical, although hemispherical ends are frequently used. The volume of the vessel is about 30 to 50% larger than the required culture volume, leaving a headspace allowing disengagement of liquid droplets from the exhaust gas and room for foaming.

The *agitator* consists of one or more impellers, mounted on a shaft, usually suspended on a thrust bearing above the vessel and entering the vessel through a gland or mechanical seal. The shaft is driven by a motor via a flexible coupling, a clutch, or, on smaller units, a belt drive or 'universal joint'. With very long shafts, one or more steadying bearings are used, to reduce vibration and wear on the shaft seal. Such bearings must be of very simple types, lubricated by the nutrient medium and capable of repeated sterilization. The gland or seal through which the shaft enters the vessel is a major contamination risk point, and must provide a tight closure between the vessel and the atmosphere, while allowing the shaft to rotate reasonably freely. The elements of a mechanical seal, illustrated in Figure 4.5, are two seal rings with accurately flat faces rotating against each other, and forming the boundary between the fermenter and the outside atmosphere. A 'soft' seal ring of carbon rotates with the shaft and is held in contact with a 'hard' seal ring of ceramic or hard-faced stainless steel by springs made of metal, rubber or plastic. The seal is cooled and lubricated by a flow of sterile liquid, usually steam condensate. Additional security against contamination is obtained by mounting two such seals 'back-to-back' to form a double mechanical seal.

The *impellers* mounted on the agitator shaft effect the dissipation of energy into the fermentation medium, and are usually of a conventional and well-documented design, such as a 6- or 8-bladed turbine or vaned disc, with a

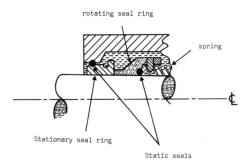

Figure 4.5   Sketch showing basic elements of a mechanical seal.

diameter conventionally one-third of the vessel diameter. The siting of impellers on the shaft is important in order to obtain the optimal flow-pattern of liquid in the vessel. Spaced too closely, multiple impellers tend to behave like a single large impeller; spaced too far apart, they leave stagnant regions in the liquid. Optimal spacing is about one impeller diameter apart, with the lowest impeller about one impeller diameter above the bottom of the vessel.

Impellers have two distinct and to some extent conflicting functions. One is to provide mixing by pumping liquid round the vessel, and the other, in sparged systems, to disperse the injected gas-stream as small bubbles and re-disperse coalesced bubbles. Generally the mixing function requires a large-diameter, low-speed impeller with a small number of blades, whereas gas dispersion requires a high-speed, small-diameter impeller with a large number of blades (Solomons, 1980), so that the conventional configuration of a stack of similar impellers mounted on a single shaft must represent a compromise between the two functions. In a design patented by Le Grys and Solomons (1977), the functions are separated by using separate, independently-driven impellers on short shafts, a low-speed 'pump' impeller at the top, and a high-speed, 18-bladed turbine impeller at the bottom. Each impeller is optimal for its function, and all but 10% of the total power input is dissipated by the high-speed turbine. Mounting the impellers on separate, short shafts means that very deep tanks can be used without correspondingly long agitator shafts. Very deep tanks give a high hydrostatic pressure at the bottom of the vessel, which enhances gas dissolution, and also enables large-capacity vessels to have diameters small enough for transport by road or railway. This means that large vessels can be fabricated in the constructor's own workshops, with the concomitant higher standard of fabrication. The use of separate shafts does, however, mean an additional shaft entry to the vessel and thus an additional major contamination risk point. This configuration was designed to deal with the very viscous broths developed in the culture of filamentous fungi.

In conventional stirred-tank fermenters, the turbulence required for satisfactory mixing, heat- and mass-transfer is obtained with the aid of four baffles attached or close to the vessel wall. The width of the baffles is usually 10–12% of the vessel diameter, as the mixing effect is increased only a little with wider baffles, but decreases sharply with narrower baffles. The agitator power dissipation in production-scale fermenters is about $1-2\,\mathrm{kW\,m^{-3}}$.

For aerobic fermentations, air is injected through a sparger, a single nozzle or a perforated tube arrangement, sited well below the lowest impeller to avoid swamping it with gas. The sparger should have provision for drainage, so that no culture medium remains in it after the vessel is discharged. The rate of air supply must be sufficient to satisfy the oxygen demand of the fermentation after allowing for the efficiency of oxygen dissolution achieved, which depends on the design and operation of the system. The practical upper limit of gas flow rate is that which can be efficiently dispersed by the impellers, which depends on their design and speed of rotation. Gas flow rates actually used are typically 1% of the culture volume per second, but subject to variation by the control system according to the growth phase of the organism and its oxygen demand. The gas sparge dissipates power and contributes to agitation of the system, but gas compression is about 40% efficient, whereas mechanical agitation is about 90% efficient (Solomons, 1980). The power dissipated by agitation has to be removed via the fermenter cooling system, but the $1-2\,\text{kW}\,\text{m}^{-3}$ is small compared to the $30\,\text{kW}\,\text{m}^{-3}$ of metabolic heat liberated by an intensive fermentation. Some designs combine the impeller and sparger, so that gas enters the system through a hollow agitator shaft and a perforated or porous impeller. The hydraulic shear then sweeps off gas as small bubbles before they can grow and detach in the normal way. These devices are used successfully in waste-water treatment with high-purity oxygen, where high oxygen utilization is essential.

Instead of a rotating stirrer, some systems effect the mechanical power input by using a *pump* to circulate liquid medium from the fermenter vessel through a gas entrainer and then back into the fermenter. This separates the liquid movement and gas dissolution functions into separate specialized units, and two designs have evolved using this principle, the 'loop' fermenter and the 'deep jet' fermenter (Figure 4.6). In the loop fermenter, the gas dissolution device is a subsidiary vessel into which gas is injected, and the gas-saturated liquid is recirculated to the main growth stage. In the deep-jet system, gas is entrained into a high-power jet of liquid injected into the liquid in the fermenter, re-entraining gas from the vessel head-space. Exhaust gas is purged partly from the vessel head-space and partly from the specially-designed circulating pump, from which the degassed liquid passes through a supplementary cooler before passing to the gas entrainer. This system gives high gas dissolution rates, but has a correspondingly higher power consumption compared to conventional systems. The liquid and entrained gas can also be introduced into the fermenter through a 'bell', which holds the gas bubbles in contact with the recirculating liquid to enhance gas utilization. Both this and the plunging jet technique are successfully used in biological waste-water treatment (Winkler, 1981).

4.2.5 *Gas-lift and sparged-tank fermenters*

In this design, there is no mechanical stirrer, and the power dissipation for mixing, heat-transfer and gas dissolution by the movement of gas through the liquid medium. The gas is thus the power transmission system from the gas compressors into the vessel. While the relatively low efficiency of gas compression seems to make this design unattractive, it has some important

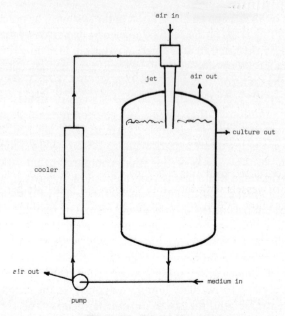

Figure 4.6 Deep-jet fermenter.

advantages compared to the stirred-tank system. Firstly, the absence of a rotating agitator shaft removes the major contamination risk at its entry point to the vessel. Secondly, with very large vessels, the required power input for agitation is just too large to be transmitted by a single agitator. For example, a vessel holding 1500 tonnes of medium—about ten times larger than a conventional stirred-tank fermenter—the requirement is about 2 MW. A mechanically-agitated system would require multiple agitators with multiple shaft entry-points and there is a correspondingly increased contamination risk. Thirdly, the evaporation of water vapour into the gas stream makes a small contribution to cooling the fermentation. The fermenter interior does, however, need careful design to ensure that the movement pattern of the gas through the system produces satisfactory agitation, and there is a large range of different designs of gas-sparged fermenters, each with particular advantages claimed for it.

Agitation cannot be increased merely by increasing the gas flow rate, as there is an upper limit beyond which quantities of liquid are carried out of the vessel in the gas stream. This is expressed in terms of the *superficial gas velocity*, which is the velocity the gas would have if it passed through the empty vessel at a uniform rate, or, mathematically, the volume gas flow-rate ($m^3 s^{-1}$) divided by the cross-sectional area of the vessel (m^2). At high superficial gas velocities, gas bubbles coalesce to form 'slugs', which not only carry liquid out of the system, but also reduce the efficiency of gas dissolution by reducing the gas–liquid interfacial area. Fitting horizontal perforated baffles in the vessel will break up the slugs and re-disperse them as bubbles, and liquid carry-over can be reduced by installing a disentrainment device before the gas outlet. This can be a head-space with a larger cross-sectional area than the main part of the vessel, with thus a lower superficial gas velocity, or a separate disentrainment device such as a cyclone, although separate stages can act as separate growth systems contaminating the main fermentation.

It should be noted that the superficial gas velocity is a scale-dependent parameter. For vessels with the same volume gas flow-rate per unit volume, the volume gas flow-rate increases proportionately to vessel volume, whereas the superficial gas velocity increases proportionately to the *cube root* of the vessel volume. This means that large vessels operate closer to the limiting superficial gas velocity for the design than smaller ones.

The various designs of non-mechanically agitated fermenters can be grouped broadly into sparged vessels and gas-lift (including air-lift) fermenters (Figure 4.7). *Sparged-tank* (or 'bubble-column') fermenters are usually of high aspect ratio, with gas introduced at the bottom through a single nozzle or a perforated or porous distributor plate. The gas bubbles rise through the liquid in the vessel and may be redispersed by a succession of horizontal perforated baffle-plates sited at intervals up the column. Temperature control is maintained by attemperator jackets or internal coils. In anaerobic digestion, the gas evolved can be recirculated through spargers to maintain homogeneity and enhance temperature control, but since gas dissolution is irrelevant, baffle-plates are not used. Similarly, the rise of bubbles of carbon dioxide evolved in brewing fermentations can be used to provide agitation in the 'conical' fermenter.

In *gas-lift* fermenters, internal liquid circulation in the vessel is achieved by sparging only part of the vessel with gas. The sparged volume has a lower effective density than the bubble-free volume, and the difference in hydrostatic pressure between the two sections drives the liquid circulation upwards in the sparged section and, after gas disentrainment, downwards in the bubble-free section. The two sections may be separated by a vertical draught-tube. Overall, the vessel acts more or less as a completely-mixed system, but within the circulation loop has the properties of a plug-flow system. The air-lift fermenter is the type of design selected for a recent development in fermentation engineering, the ICI 'Pruteen' bacterial SCP process (see Figure 4.9). The liquid height of 45 m in the process gives a high hydrostatic pressure at the bottom of the vessel, which enhances air dissolution. The aerated liquid circulates upwards through a succession of

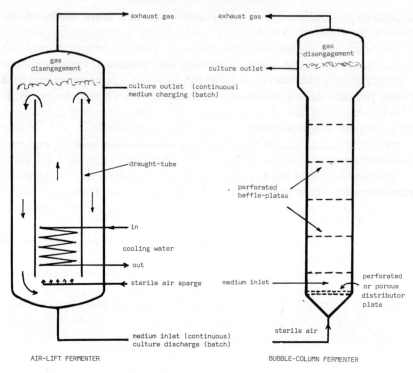

Figure 4.7 Gas-sparged fermenters.

perforated baffles to ensure uniform power dissipation and to redisperse air bubbles. The lower pressure at the liquid surface and the larger cross-sectional area at the top of the column enhance disentrainment of unutilized gas and carbon dioxide evolved in the process. As the growth medium circulates several times in passing through the vessel, it undergoes successive cycles of high and low pressure, so the system is known as the 'pressure-cycle' fermenter. A problem occurring with fermenters with a high gas sparge-rate is the shear effect due to the high velocity of the gas leaving the distributor, which has been found to damage certain types of organism.

4.3 Heat transfer

The heat transfer duties of a fermenter are to maintain a specified temperature during the course of a fermentation, and to effect rapid and complete heat sterilization. These two duties involve different types of performance. Heat sterilization requires rapid heating to, holding at, and cooling from temperatures well over 100°C. In fermentation, however, temperature control requires the continual removal of metabolic heat at a fairly steady rate to maintain temperatures little different from ambient, usually around 30°C. Of the two, temperature control is more critical. Failure to maintain the required temperature during fermentation can lead to low yield, loss of product or even the wrong product. In heat sterilization, the use of the wrong temperature can usually be corrected by modifying the holding time at that temperature, providing it is actually high enough to effect sterilization, although this may incur expense and inconvenience.

Transfer processes can be described by an equation in which the rate of transfer of a property—heat or mass—is proportional to the area available for the transfer process and to the driving force for transfer. The 'constant' of proportionality is then the 'transfer coefficient' for the property:

$$\text{transfer rate} = \text{transfer coefficient} \times \text{area} \times \text{driving force} \quad (4.11)$$

For heat transfer, the driving force is the temperature difference between the two systems between which heat is transferred, so that for a temperature difference $\Delta T(K)$ across an area $A(m^2)$ with a heat-transfer coefficient U (W m^{-2} K^{-1}), the heat transfer rate, Q (W) will be

$$Q = U \cdot A \cdot \Delta T \quad (4.12)$$

For heat transfer into a system of volume $V(m^3)$,

$$Q/V = U \cdot \frac{A}{V} \cdot \Delta T \quad (4.13)$$

or if q is the heat transfer rate per unit volume (W m^{-3}) and a the 'specific surface' or surface-area per unit volume (m$^2 \cdot$ m^{-3}), then

$$q = U \cdot a \cdot \Delta T. \quad (4.14)$$

Equation (4.14) demonstrates that for a given system, the *heat transfer rate per unit volume depends on the specific surface of the system, the temperature difference and the heat-transfer coefficient obtainable*. These parameters may not easily be quantifiable, for example when a liquid is heated by direct steam sparging, and may vary throughout a system. For example, the temperature difference changes as the hot side cools and the cool side is warmed up, and the heat-transfer coefficient is affected by the temperature of the liquid and the intensity of agitation. In some cases, useful approximations can be made by taking a suitable average value, such as a logarithmic mean temperature difference.

Fermentation liberates heat and most industrial fermenters are fitted with *cooling systems*. Cooling water can be circulated through jackets surrounding the vessel, or through coils inside the vessel immersed in the culture, or sprayed on to the outside surface of the vessel. In the last case, evaporation of the water makes an important contribution to the cooling effect, but should be used only in a well-ventilated environment. A closed environment will quickly become saturated with water vapour, which will create unpleasant working conditions, encourage microbial growth and also suppress the evaporative cooling effect. With jackets and coils, temperature is controlled by removal of sensible heat by the cooling water, whose flow is regulated by pumps or valves actuated by the temperature control system. If the cooling water flow-rate increases, the rate of heat transfer increases, because the cooling water is warmed up less, and the temperature difference between it and the culture is increased. In addition, the extra turbulence at higher flow-rates tends to increase the overall heat-transfer coefficient.

Cooling jackets leave the fermenter interior uncluttered but have a lower heat-transfer coefficient than coils, and the relative surface area (a in eqn 4.14) is scale-dependent and, for similar vessels, is inversely proportional to the vessel diameter. Internal coils have a relative heat-transfer surface virtually independent of scale, as larger vessels have more space available for coils, and have a higher heat-transfer coefficient. On the other hand, coils take up space inside the vessel, provide additional surface for the adhesion of microorganisms, are difficult to clean and can interfere with gas dispersion in gas-sparged systems. Where the cost can be justified, a vessel can be fitted with both internal coils and an external jacket. One type of jacket is in the form of a coil welded to the outside of the vessel shell, which increases the turbulence and heat-transfer coefficient in the coolant flow. External jackets, in spite of their limitations, are preferred in some designs because of the 'clean' vessel interior they give, and the required heat transfer rate is maintained by using a chilled coolant—chilled water or alcohol solution—so that the increase in ΔT in eqn 4.14 compensates for the lower values of a and U.

In general, the effect of scale on the relative heat transfer area means that jackets are suited to smaller vessels and coils to larger vessels. The break-even point depends on the geometry of the vessels and the rheology of the growth medium, but, taking typical values, Richards (1968) showed that jackets and coils give similar specific heat transfer rates with vessels of diameter of about 3 m, the diameter of a typical production-scale stirred-tank fermenter. Values for the metabolic heat load cited by Smith (1980) are about $30\,\text{kW}\,\text{m}^{-3}$ for yeast growing on carbohydrate and $65\,\text{kW}\,\text{m}^{-3}$ for yeast growing on hydrocarbon. Added to this is the agitator power dissipation of about $1-2\,\text{kW}\,\text{m}^{-3}$. Evaporation of water vapour into the sparged air flow can remove roughly $0.5\,\text{kW}\,\text{m}^{-3}$.

Similar principles apply to processes running at relatively high temperatures, which can be heated with hot water from another process, such as steam condensate or cooling water from a generator motor.

4.3.1 Heat sterilization

Before starting a fermentation, the growth medium and the plant and equipment involved are sterilized, so that the desired organism can be introduced and grown free of contamination and competition. The usual and most convenient method of sterilization is by heating to a temperature high enough to kill living organisms, holding at that temperature long enough to achieve sterility and then cooling to culture temperature. The rate at which organisms are killed increases rapidly with temperature, so the higher the temperature used, the shorter the holding time needed. The time required at any temperature is calculated by standard methods based on the thermal inactivation of the heat-resistant spores of a standard test organism. For details see Richards (1968) or Banks (1979).

Two alternative sterilization procedures are available. The fermenter vessel can be charged with all or part of the growth medium and sterilized together, or the vessel and the medium can be sterilized separately, with the sterile medium then charged aseptically to the sterile vessel. *Separate sterilization* is obligatory for continuous processes, and has the advantage that sterilization of the medium can be carried out in a specifically-designed unit providing sterile medium for several fermenters. The fermenters themselves are then sterilized empty, which takes up less of the cycle time than sterilizing a fully-charged vessel, thus increasing fermenter productivity. It is in any case useful to be able to sterilize at least part of the medium separately, as some components can react together at sterilization temperatures and reduce the nutritional value of the medium, for example Maillard reactions between amino acids and sugars. The nitrogenous components can be sterilized separately and added aseptically after cooling to growth temperature. The capacity of the separate sterilizing unit can be reduced by sterilizing the medium in a concentrated form and then diluting it in the fermenter with sterile water. The disadvantage of separate sterilization is the risk of contamination during transfer between the sterilizing unit and the fermenter. *Sterilization of the fermenter and its contents together* is suited only to batch operation and maintains the vessel virtually as a closed system, but contact between the vessel and the medium at elevated sterilization temperatures may cause corrosion or deposition of pyrolized nutrients on the vessel surfaces. Provision must be made for the venting of gas from the vessel during heating and the admission of sterile gas on contraction during cooling. For batch processes, the choice between separate and *in-situ* sterilization depends on the flexibility of operation required. The cost of a specially-designed sterilization unit is roughly the same s that of a fermenter and can be justified only when one sterilizer serves several fermenters. An operation with only one or two fermenters would probably benefit more from an additional fermenter than a separate sterilizer. A hold-up in a separate sterilizer serving several vessels will affect all the fermenters, whereas with *in-situ* sterilization, each fermenter is independent of the others.

The *heating phase* can be carried out as a batch process in a specially-designed agitated cooker, or in the fermenter itself, by passing steam into the coils or jacket, or, with a suitable medium, by direct sparging of steam into the liquid. Steam sparging is very rapid, as there is no solid heat-transfer barrier between the steam and the medium, but in most growth media produces severe foaming. The steam must be free of anti-corrosion additives, and allowance must be made for dilution by the steam condensate. Continuous heating is carried out after compounding the medium in a non-sterile mixing vessel by passing it through plate or spiral heat exchangers or a venturi steam injection device. The hot medium is held at temperature by passing through a well-lagged tube where the residence time is sufficient to effect sterilization at that temperature, and is then cooled in heat exchangers or a flash evaporator. With heat exchangers, heat economy is obtained by using the hot sterile medium to heat up the incoming raw medium. Flash evaporation gives virtually instantaneous cooling but can also produce severe foaming. The time taken by a sterilization cycle depends on the temperature and equipment used. With standard equipment and sterilization at 121°C—the temperature of saturated steam at one atmosphere—the cycle time could be two or three hours. With very rapid heating to and cooling from, say, 150°C in a continuous system, the cycle-time could be only a few minutes.

Sterilization of empty vessels is commonly carried out by direct sparging with wet steam, which gives much more rapid sterilization than dry saturated or superheated steam at a given temperature. Air must be expelled from the vessel, as air pockets can blanket parts of the vessel from contact with the steam, and air also reduces the partial pressure of steam in the vessel. After the appropriate holding time at sterilization temperature and pressure, the vessel is cooled using the normal cooling system, assisted by a flow of sterile air through the vessel. The vessel must be thoroughly cleaned before sterilization as a dirt layer protects the vessel surface from contact with the steam. Large vessels with a high heat capacity will produce large volumes of steam condensate, and this must be cleared through efficient steam traps so that it does not prevent access of steam to the lower regions of the vessel.

4.3.2 *Sterilization without heat*

Alternative sterilization methods must be used for materials which are liable to damage by heat sterilization. Nutrient media may contain heat-sensitive components or undergo reactions at sterilization temperatures, and some plant components may be affected by the temperature and pressure reached in heat sterilization or by the repeated heating and cooling of the sterilization cycle.

Liquids free of suspended solids can be sterilized by passing through absolute filters with a pore size small enough to trap any organism in the liquid. Examples of such filter media are unglazed ceramic and plastic membranes, which are available with a selection of pore sizes down to those small enough to trap small virus particles, and granular precoat filters, which remove bacteria. Preliminary filtration of the liquid with a coarse granular or fibre medium is advisable, as apparently clear liquids can contain particles that will rapidly block ('blind') a sterilizing filter. Sterile filtration removes all particles above the designated size—organic, inorganic, living or dead—whereas heat sterilization leaves the liquid free of *living* organisms but not of the resultant biological debris. This is important in the production of pyrogen-free water for pharmaceutical use.

The main problem of sterile filtration is that the flow rate for a given pressure difference falls sharply as the pore size used is reduced, so that very large filter surface areas are needed to obtain industrial-scale flow rates. The volume of non-heat-sterilized liquid is thus kept to a minimum, with heat-sensitive soluble materials sterilized as a concentrate by filtration and added aseptically to the heat-sterilized bulk of the medium.

Heat-sensitive materials containing suspended solids, which cannot be sterilized either by heating or filtration, can be sterilized by radiation, or by treatment with a chemical sterilizing agent such as ethylene oxide or β-propiolactone. Chemical sterilization can also be used for plant components likely to be damaged by the temperatures and pressures used in heat sterilization, such as non-metallic vessels and components, delicate probes and sensors used in process monitoring, and composite components made up of materials with widely different coefficients of thermal expansion. Thorough cleaning is needed before treatment to ensure efficient contact between the biocidal chemical and the component; some surface-active materials combine the biocidal effect with the ability to loosen and remove deposits from surfaces. Good design and construction of equipment is particularly important where chemical disinfection is used, to avoid crevices and rough areas where microorganisms can lodge or restrict access to a biocidal fluid. Different biocidal fluids have different effects on bacteria, fungi, spores or viruses, on the plant components and on operating staff, which must be taken into account. Residual disinfectant must be removed after treatment by washing with sterile water, and effluent treatment problems may also be caused if large quantities of biocide are discharged to a biological waste-water treatment plant. Examples of disinfecting solutions are 10% hypochlorite, after treatment with which residual chlorine is removed by rinsing with a solution of sodium thiosulphate in sterile water, and quaternary ammonium compounds (QAC), which are bactericidal detergents. Fumigation with gaseous or volatile biocides, such as ozone, formaldehyde, sulphur dioxide, ethylene oxide or β-propiolactone is also used. These have disadvantages: formaldehyde and ethylene oxide tend to leave a polymeric coating on surfaces that can be difficult to remove, sulphur dioxide is corrosive, ethylene oxide explosive and β-propiolactone is reported as being carcinogenic.

4.4 Mixing

4.4.1 *Introduction*

Mixing is required to ensure that the desired growth environment is obtained throughout a process, as deviations from optimal conditions result in impaired yields. Mixing is obtained by creating turbulence through agitation, using mechanical stirring and/or gas sparging. The resulting forced convection also enhances heat-transfer (for temperature control) and mass-transfer (for gas dissolution) and rapidly disperses injected control fluids.

The degree of turbulence induced in a fluid by agitation is expressed by the *Reynolds number* (*Re*) of the system, and, provided consistent units are used in its calculation, it is dimensionless. The Reynolds number is thus a 'dimensionless group' and relates the inertial and viscous forces acting in agitation. Its general definition is

$$Re = \frac{D \cdot u \cdot \rho}{\eta} \qquad (4.15)$$

where D is a characteristic linear dimension of the system, such as a pipe or particle diameter, u a characteristic velocity, such as that of a fluid or a particle, and ρ and η the density and viscosity of the fluid involved respectively. The different regimes of fluid flow are described by the same values of the Reynolds number for several different systems, and for *fully-developed turbulence, Re needs to be greater than about 10^4*. For liquid in a mechanically-stirred tank, the appropriate quantities in calculating the Reynolds number are the diameter and blade-tip velocity of the agitator impeller, with the density and viscosity of the liquid, so that

$$Re = N \cdot D_i^2 \cdot \rho/\eta \qquad (4.16)$$

where D_i is the impeller diameter and N its speed of rotation. The problem with many fermentations is that the viscosity η is not constant but is a function of the growth phase of the organism, and the impeller diameter and speed (this is discussed further in section 4.4.2).

The intensity of agitation can be expressed in terms of the power dissipation into the liquid, P, which can be related to the system characteristics in a dimensionless group called the *power number* N_p, defined by

$$N_p = P/\rho \cdot N^3 \cdot D_i^5 \qquad (4.17)$$

In an agitated tank where the fluid flow is not fully turbulent, the power number varies with the Reynolds number, but with fully-developed turbulence, the power number reaches a *constant value independent of the Reynolds number*. The actual value of N_p achieved depends on the geometry and configuration of the system, but for a standard fully-baffled tank with a six-bladed turbine impeller of diameter one-third of the tank diameter, it is about 4–6. The effectiveness of mixing can be expressed in terms of a mixing time, t_m, the time taken to disperse a liquid droplet completely into liquid of the same physical properties. In practice, the mixing time is taken as the time required to disperse a specific factor to a predetermined concentration level. Norwood and Metzner (1960) defined a mixing-time factor, Φ_m, which, for geometrically similar vessels, was related to the mixing time in an expression

$$\Phi_m = \text{constant} \times t_m \cdot N^{2/3}/D_i^{1/6} \tag{4.18}$$

This factor was found to vary with the Reynolds number in a manner very similar to that of the power number, and with fully-developed turbulence achieved the same value (about 6) for the standard configuration of vessel and impeller. Since the power number is constant in the region of fully-developed turbulence, then from equation 4.17, the power dissipation, P, is given by

$$P = \text{constant} \times N^3 \cdot D_i^5 \tag{4.19}$$

For geometrically similar vessels, the volume, V, is proportional to the cube of the impeller diameter, so for vessels with a constant power input per unit volume,

$$P/V = \text{constant} \times N^3 \cdot D_i^2 \tag{4.20}$$

Combining equations 4.18 and 4.20, for fully-turbulent, geometrically similar vessels with the same power input per unit volume, the mixing time t_m is proportional to the impeller diameter (or any other characteristic linear dimension of the vessel) to an exponent of 11/18, or approximately $D_i^{2/3}$, and thus to the square of the vessel volume. This implies that mixing deteriorates rapidly as the volume of a particular design of vessel increases. While this is found in practice to some extent, the effect is much less dramatic. This can be shown by considering the change in power input per unit volume needed to maintain a constant mixing time in geometrically similar vessels. With fully-developed turbulence giving a constant value of Φ_m, then, from eqn 4.18 for constant mixing time

$$N = \text{constant} \times D_i^{1/4} \tag{4.21}$$

and from eqn 4.20, for constant mixing time,

$$P/V = \text{constant} \times D_i^{11/4} \tag{4.22}$$

and since the vessel volume is proportional to D_i^3, then

$$P/V = \text{constant} \times V^{11/12} \simeq \text{const.} \times V \qquad (4.23)$$

This implies that, insofar as mixing conditions are described by t_m, then the power input *per unit volume* must be increased proportionally to vessel volume to maintain the same mixing conditions. Operational experience shows that power input per unit volume should actually be reduced with increasing vessel size to maintain equivalent mass-transfer conditions. This shows that, although mixing is important, it may be overridden by other considerations. Mixing with gas sparging alone is much less fully documented, but in general increases with the superficial gas velocity up to a maximum limiting value.

4.4.2 *Outline of rheology*

Fluid mixing is carried out by applying forces, through agitator impellers or gas sparging, that induce fluid movement. The rheology of a fluid describes the way it reacts to these forces and its resultant movement. In fermentation broths, the rheological properties may be strongly affected by the presence of certain types of medium components and of the microbial culture in it. One of the problems in fermenter design is that this behaviour is not only difficult to predict, but may change during the course of a batch fermentation. The components of a broth can confer quasi-solid properties on the liquid medium in addition to its inherent liquid properties.

When a shear-force (force per unit area) is applied to a solid, the solid is deformed but tends to resume its original shape when the shear-force is removed. The solid is then said to have shape-retaining or 'elastic' properties. When a shear-stress is applied to a fluid, it produces a velocity gradient or 'shear-rate' in the fluid, and the ratio of the shear-stress, τ, to the shear-rate, γ, is called the *apparent viscosity*, η_a, of the fluid, so that

$$\eta_a = \tau/\gamma \qquad (4.24)$$

Fluids having elastic or shape-retaining properties as well as viscous shear-resisting properties are called 'viscoelastic fluids' and are discussed later. Fluids whose elastic properties are negligible are described by the relation between shear-stress and shear-rate in the fluid, which can be generalized as

$$\eta_a = f(\gamma, t) \qquad (4.25)$$

where t is time, so that the apparent viscosity is a function of the rate and duration of shear.

Fluids encountered in fermentation show several different behaviour patterns, each of which can be considered as a special case of eqn 4.25. Fluids where the viscosity is *independent* of shear-rate are known as *Newtonian*

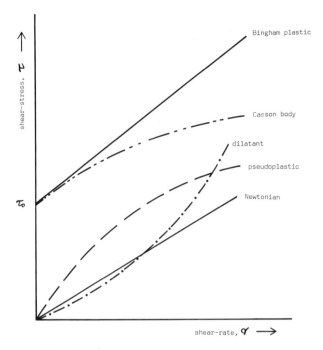

Figure 4.8 'Time-independent' non-Newtonian fluids.

fluids, so the others are called *non-Newtonian* fluids. Some fluids contain materials which can form or lose a structure under the influence of shear, and the apparent viscosity of the fluid depends on this structure and thus on the shear-rate. At constant shear-rate, the viscosity changes until equilibrium is established between the effects of shear and the tendency of the fluid to assume its former structure. As this takes time, such fluids are called *time-dependent non-Newtonian* fluids, and their behaviour exhibits hysteresis, such that it depends on its approach to equilibrium. Such fluids whose inherent structure tends to be disrupted by shear and whose viscosity falls with time towards equilibrium are called 'thixotropic' (an example is non-drip paints). Those in which shear induces the formation of a structure and whose viscosity increases with time towards equilibrium are called 'rheopectic', for example plaster of Paris and dilute slurries of bentonite. Fluids whose structure changes so rapidly with time that the effects of shear can be taken as virtually instantaneous are called *time-independent non-Newtonian* fluids, and are classified according to the relation between shear-stress and shear-rate in the fluid (Figure 4.8).

Fluids such as toothpaste, drilling muds and sewage sludge have at rest a structure which resists shear and must be disrupted before fluid flow commences. This means they have a definite yield stress, τ_0, so that

$$\tau = \tau_0 + \zeta\gamma \tag{4.26}$$

where τ_0 is the yield stress and ζ the *plastic viscosity* or *coefficient of rigidity*. This type of fluid is called a 'Bingham plastic', and several fungal fermentation broths have been reported as having this rheology.

Fluids containing filamentous components show a decrease in apparent viscosity with increasing shear-rate as the filaments become progressively aligned along the direction of shear, so reducing the resistance to shear. Some fungal fermentation broths behave in this fashion, as the mycelial filaments align with the direction of shear, as do the molecules in some polymer solutions, and these are called 'pseudoplastic' fluids. Printing inks and wallpaper paste are other examples, designed to spread easily under high shear, but not to drip under low shear. The relation between shear stress and shear rate is

$$\tau = K\gamma^n \tag{4.27}$$

where K is called the *consistency coefficient* and n the *flow behaviour index*, which for pseudoplastic fluids is less than one, although several other equations have been developed. In high-concentration solid suspensions, the apparent viscosity increases with the shear rate so that the flow behaviour index n, in eqn 4.27 is greater than one. This is explained by assuming that, at low shear rates, the solid particles remain apart from each other so that their movement is lubricated by the fluid between the particles. At high shear rates, particles are pushed together, so removing the fluid lubricant and making movement more difficult. These are called 'dilatant' fluids, for example wet cement. Most solutions not containing particles or polymers are Newtonian fluids showing a constant viscosity or shear stress/shear rate ratio. This viscosity is increased by the presence of suspended particles, for example small, single-celled organisms such as yeast, described by the Einstein equation for dilute suspensions

$$\eta_a = \eta(1 + 2.5\varepsilon) \tag{4.28}$$

where η is the Newtonian viscosity of the suspending liquid and ε the volume fraction occupied by the particles. For more concentrated suspensions of small spheres, with ε up to 0.25, the Vand equation applies:

$$\eta_a = \eta(1 + 2.5\varepsilon + 7.5\varepsilon^2) \tag{4.29}$$

A type of non-Newtonian fluid that has a yield-stress τ_0 and whose apparent viscosity falls with increasing shear-rate is called a Casson body, described by

$$\tau^{1/2} = \tau_0^{1/2} + K_c \cdot \gamma^{1/2} \qquad (4.30)$$

Some fungal fermentation broths behave in this manner.

Viscoelastic fluids contain materials that confer elastic or shape-retaining properties on the fluid in addition to shear-resisting viscous properties, so that when its motion stops, the stress relaxes in an exponential decay pattern. A relation describing these will include an elasticity modulus as well as viscosity terms, and elastic deformation involves forces normal to the direction of shear as well as parallel to it. This leads to unusual effects such as the formation of a 'vena expanda' in fluid jets, the Weissenberg effect where polymer melts climb up the agitator shaft on stirring, and the reversal of flow patterns on stirring under some conditions. Examples of viscoelastic fluids are blood, egg-white, thick soup and molten synthetic polymers; Bingham plastics can also be thought of as a special case of a viscoelastic fluid with zero elasticity.

When polymeric substances, such as polysaccharide nutrients or synthetic products of the organism, are introduced into a Newtonian solution, its viscosity will increase and its rheology change to non-Newtonian. When single-celled organisms, such as bacteria or yeast, are grown in a Newtonian medium, then Newtonian characteristics are retained, although the viscosity increases according to the Einstein or Vand equations (4.28 and 4.29). When an organism forms filamentous growths or aggregates into deformable clumps, the rheology becomes non-Newtonian. Broths containing short filaments have lower apparent viscosities than those of the same biomass content with long filaments, although if the filaments roll up to form pellets, the viscosity falls and becomes almost Newtonian. Some batch cultures of filamentous organisms change their rheology during the fermentation, with viscosity and non-Newtonian behaviour increasing at first as filaments grow, then changing as the filaments break up or form pellets. For example, cultures of *Streptomyces* have been reported as starting and ending with Newtonian behaviour, passing through an intermediate Bingham or pseudoplastic phase. *Penicillium chrysogenum* in penicillin manufacture has been reported as a Casson body, Bingham plastic and pseudoplastic. Bingham and pseudo-plastics can be confused when there is insufficient data at low shear-rates to show the presence or otherwise of a yield stress.

In trying to calculate the power requirements for agitation of non-Newtonian fluids, two problems are encountered. Firstly, the shear rate and thus the viscosity is not uniform throughout the fermenter, as it is highest near the impellers, and secondly, some fermentation broths are so viscous that it is virtually impossible to achieve fully-developed turbulence, so that the power number is not independent of the Reynolds number, however that may be defined. Metzner and Otto (1957) used an average shear-rate, $\bar{\gamma}$, assumed to be proportional to the impeller speed, N, which was assumed to give an average apparent viscosity, $\bar{\eta}_a$, for the vessel, so that

$$\bar{\gamma} = k \cdot N \tag{4.31}$$

and for a 'power-law' fluid following eqn 4.27 for pseudoplastic or dilatant fluids,

$$\bar{\eta}_a = K(k \cdot N)^{n-1} \tag{4.32}$$

They determined the power number/Reynolds number curve for a particular vessel using Newtonian liquids, and repeated the procedure with 'power-law' non-Newtonian fluids at different impeller speeds in the region of viscous flow, and determined the average apparent viscosity by assuming that the Newtonian and non-Newtonian Reynolds numbers were the same for the same power number. This gave a value of k of about 13; later work with a wide range of conditions gave k as about 11. Calderbank and Moo-Young (1959) developed a modified Reynolds number Re^*, for 'power-law' fluids, where

$$Re^* = \frac{D_i^2 \cdot N^{(2-n)} \rho}{K} \cdot 2^{(3-n)} \left\{ \frac{n}{3n+1} \right\}^n \tag{4.33}$$

which, when plotted against the power number, gives a curve with a shape similar to that given by Newtonian fluids, reaching a constant value of N_p for Re^* greater than about 10^4, and where n and K are the same as in eqn 4.27.

4.5 Oxygen supply in fermenters

The main problem of oxygen supply to fermentation is that oxygen is only sparingly soluble, so that even in batch fermentations, it must be supplied continuously. A cubic metre of culture medium will hold 7–8 g of oxygen, which will last an intensive yeast culture only a few seconds at oxygen uptake rates of 2–6 g m^{-3} s^{-1}. The process of making oxygen available to a growing culture involves bringing the gas into contact with the liquid, dissolving the gas into the liquid and then transferring the dissolved gas from the gas–liquid interface to the organisms. Rapid oxygen supply thus needs a large area of contact between gas and liquid to facilitate dissolution and turbulence and mix the dissolved gas into the bulk of the fermentation. Thus, in aerated systems oxygen supply and agitation are virtually inseparable.

These effects can be summarized by an equation based on the same principle as eqn 4.11 and analogous to eqn 4.14, for the *oxygen transfer rate* (OTR) in mass of oxygen per unit volume of culture per unit time, or g m^{-3} s^{-1} as

$$\text{OTR} = k_L \cdot a \cdot \Delta C \tag{4.34}$$

In this case, it is very difficult to distinguish the separate effects of creation of gas–liquid interfacial area per unit volume (a) from those of increasing the mass-transfer coefficient (k_L), so that the product $k_L \cdot a$ tends to be considered as a single parameter. There is no universally accepted terminology for $k_L \cdot a$, but to avoid confusion it is probably best referred to as the *volumetric* or *overall mass-transfer parameter*. The driving force in this case is the difference in dissolved oxygen concentration between the gas–liquid interface and the bulk of the liquid. The interfacial concentration is usually assumed to be the equilibrium or saturation value C^* for the gas and the liquid under those conditions, so that eqn 4.34 becomes

$$\text{OTR} = k_L \cdot a(C^* - C_L) \tag{4.35}$$

where C_L is the bulk liquid concentration.

From this it can be seen that the *OTR is increased by increasing $k_L \cdot a$ and C^* and by decreasing C_L*. The value of the bulk liquid concentration is important, because at low values it may inhibit the growth of an aerobic culture due to lack of oxygen availability, while at high values close to the interfacial concentration, the oxygen-transfer rate is impaired. The value of the equilibrium or interfacial concentration, C^*, is a function of the partial pressure of oxygen in the gas phase in contact with the liquid. This in turn depends on the fraction of oxygen in the gas phase and its total pressure. Considering a bubble of air rising through a culture, the fraction of oxygen and total pressure on entering the bottom of the vessel will both be high, and both will fall as the bubble rises due to dissolution of oxygen from the gas, stripping of product gases such as carbon dioxide from solution into the bubble and the fall in hydrostatic pressure. Where details of the change in interfacial oxygen concentration cannot be obtained, then for calculation purposes the logarithmic mean of the inlet and exhaust values is usually used. The saturation or equilibrium value C^* of the dissolved oxygen concentration is also strongly affected by temperature, as it decreases with increasing temperature, and is also reduced by the presence of other solutes.

The value of the overall oxygen-transfer parameter, $k_L \cdot a$, is affected by the intensity of agitation, expressed as the power input per unit volume of liquid, the superficial gas velocity through the tank, the materials present in the liquid and the temperature of the system. A number of correlations between $k_L \cdot a$, the power input per unit volume and the superficial gas velocity have been reported in the general form

$$k_L \cdot a = R \cdot (P/V)^x \cdot (v_s)^y \tag{4.36}$$

where R is an empirical coefficient characteristic of the system, P the power input, V the liquid volume, and x and y are empirical exponents. Reviews of such correlations have been presented by Winkler (1981) and Banks

(1979, 1977), and show that the reported values of x and y are all less than one. This means that, while $k_L \cdot a$ can be increased by increasing, say, the power input, the gain in $k_L \cdot a$ is proportionately less than the increase in power. For example, x is typically about 0.6, so that to double the value of $k_L \cdot a$, the power input must be increased by a factor of 3.2. In systems with both mechanical agitation and gas sparging, there is a choice of methods of increasing $k_L \cdot a$, and if, as is typical, y is greater than x, then it may be advantageous to increase the superficial gas velocity rather than the power dissipation, after allowing for the greater efficiency of mechanical agitation. The mechanical power dissipation is also affected by the gas-flow in the system, as the presence of gas bubbles in the liquid reduces its effective density and thus the power input, in accordance with eqn 4.17. An empirical correlation between the gassed and ungassed power dissipations has been presented by Michel and Miller (1962) which is found to apply in a wide range of conditions:

$$P_g = R'(P_u^2 \cdot N \cdot D_i^3 / G^{0.56})^{0.45} \tag{4.37}$$

where P_g and P_u are the gassed and ungassed power consumptions, and G the volumetric gas flow rate. This can be expressed in an approximate form

$$P_g/P_u \simeq R''(G^{0.4}/N_a)^{1/2} \tag{4.38}$$

where N_a is a dimensionless 'aeration number' defined as

$$N_a = G/ND_i^3 \tag{4.39}$$

In gas-sparged vessels, the value of $k_L \cdot a$ increases steadily with increasing superficial gas velocity until bubble coalescence reduces the gas–liquid interfacial area by increasing the average bubble size. Schügerl et al. (1977) expressed this in the form

$$k_L \cdot a = 0.0023 \, (v_s/d_B)^{1.58} \tag{4.40}$$

where d_B is the mean bubble diameter. This corresponds fairly well with eqn 4.36 if x and y are taken as about 0.6 and 0.75 respectively, the bubble diameter proportional to the superficial gas velocity to an exponent of about 0.15, and the power dissipation roughly proportional to the superficial gas velocity.

The presence of dissolved salts in the liquid inhibits bubble coalescence, and $k_L \cdot a$ increases with temperature as well as increasing ionic strength, although these two parameters are usually determined by the growth requirements of the organism. The microbial culture itself has a considerable effect on the mass-transfer behaviour of the system, and in fungal fermentations $k_L \cdot a$ is severely reduced by high mycelial concentrations. The Michel and Miller correlation (eqn 4.37) is found to describe the power uptake of gassed non-Newtonian cultures reasonably well, although it is found in practice that the gassed power consumption actually decreases during the course of a batch mould fermentation, and may be reduced to as little as 30% of the initial value (Banks, 1977). This is due to the effect of the mould mycelium on the distribution of gas bubbles in the liquid, which becomes progressively less uniform as the culture becomes more viscous and the bubbles remain close to the agitator as they rise through the liquid. The agitator is thus rotating in liquid whose effective density is progressively reduced. The high viscosity of mould cultures also inhibits bubble coalescence, and this was the explanation advanced for the observation of Steel and Maxon (1962) that, in a culture of *Streptomyces niveus* producing novobiocin, greater power input was needed with a large impeller than with a small impeller to produce a given yield of novobiocin. While $k_L \cdot a$ was found to be proportional to the gassed power input to an exponent of 0.46, the coefficient of proportionality was higher for small impellers than for large impellers. A good correlation was obtained between the oxygen transfer-rate and the impeller tip speed, of the form

$$\text{OTR} = \text{constant} \times (N \cdot D_i)^{1.6} \qquad (4.41)$$

The higher shear at the tip of a small, high-speed impeller with the same power input as a large, slow-speed impeller produces smaller bubbles, whose coalescence is then inhibited by the high viscosity culture. In low-viscosity systems, much of the agitator power input goes into redispersing coalesced bubbles.

The effect of bubble size on oxygen transfer in aeration is the resultant of several different phenomena. For a given volume of gas, small bubbles give a greater gas–liquid interfacial area, which enhances mass-transfer—it increases a in eqn 4.35. Small bubbles rise more slowly through the liquid and so spend longer in contact with the liquid, which also enhances mass-transfer. The slow rise-rate creates less turbulence, however, which reduces mass-transfer—a reduction in k_L in eqn 4.35. Roughly, the reduction in k_L balances out the increases in a, so the net effect is that mass-transfer is enhanced by smaller bubbles, as, for a given gas flow-rate, there is more gas in contact with the liquid—gas 'hold-up'—at any moment.

The overall effects of fermenter operating conditions on the oxygen transfer-rate can be considered in terms of eqn 4.35. The saturation concentration, C^*, can be increased by increasing the total pressure of the system, using deep vessels with a high hydrostatic pressure at the bottom, as well as a positive pressure in the vessel head space. This will increase the OTR at the expense of air compression costs. The working level of dissolved oxygen concentration should be as low as possible while avoiding oxygen starvation in any part of the vessel. Some operators believe in working with a fairly high

DO level to act as a buffer or reserve, but the consequent reduction in OTR must be compensated by increasing $k_L \cdot a$ and/or C^*, which incurs additional power costs. As pointed out earlier, $k_L \cdot a$ can be increased by increasing the power input and/or gas sparge-rate, although the increase in OTR obtained will not be in proportion. Increasing C^* by enriching the gas with high-purity oxygen is rarely practised in fermentation, but is being increasingly used in waste-water treatment processes. The nitrogen content of air used for aeration is considered to enhance the removal of gaseous and inhibitory metabolic products, notably carbon dioxide, from the culture, and even in waste-water treatment, inhibition of some removal processes has been attributed to the depression of pH in enclosed systems by accumulated carbon dioxide (Winkler, 1981).

4.6 Scale-up in biotechnology

4.6.1 *The problems of large-scale operation*

The preceding sections provide examples of the two broad groups of problems associated with large-scale operation. 'Bulk' problems from handling large quantities of material are illustrated by the long time periods required for discharge, cleaning, sterilization and recharging fermenters. The solutions to such problems are frequently available in the form of specialized units of equipment, such as separate or continuous sterilization systems, more powerful pumps and so on, but their cost must be justified by the resultant benefit conferred. Other problems may arise from the different materials used in production-scale operations—commercial-grade chemicals rather than analytical grade, or metal process plant rather than glass, which may introduce contamination problems.

'Scale-up' problems arise from the different ways in which process parameters are affected by the size of the unit. A simple example of this is the effect of scale on the specific surface of a unit. A series of vessels of geometrically similar proportions but different volumes have their volume proportional to the *cube* of the vessel diameter, but the vessel wall area proportional to the *square* of the vessel diameter, so that the specific surface, relevant to heat transfer into cooling jackets, is proportional to the reciprocal of the vessel diameter. Microorganisms provide a dramatic illustration of the same principle, as one of the features of their success as microscopic reaction systems is their enormous surface area relative to their volume. A bacterium has a volume of, typically, $5 \times 10^{-19}\,m^3$ and a surface area of about $3 \times 10^{-12}\,m^2$, so that it has a specific area of about $6 \times 10^6\,m^2 \cdot m^{-3}$, whereas a cubic metre of water could be contained in a tank with a surface area of $6\,m^2$. As it is virtually impossible to reproduce identical environments on different scales, the problem resolves into one of identifying the key parameters involved in a process and selecting the one to which the process is most sensitive. This parameter can then be made the 'criterion of scale-up', and systems designed to maintain this key parameter at the same value at all scales of operation.

4.6.2 Selection of scale-up criteria

It is generally assumed that several environmental parameters can be maintained at desired values regardless of scale, such as nutrient availability, pH and temperature. In aerobic fermentations, the most difficult problem is that of maintaining availability of dissolved oxygen, so that this becomes the key environmental control parameter. The oxygen demand of the culture needs to be satisfied by the oxygen transfer-rate (OTR) attained in the fermenter. However, the OTR is thus dependent on the growth-stage of the culture, and, from eqn 4.35, on the prevailing level of dissolved oxygen concentration. The parameter $k_L \cdot a$ is *independent of the DO concentration* and is determined by the equipment and its operation, and is thus a more generally applicable as a scale-up criterion. From the discussion of the correlations for $k_L \cdot a$ in section 4.5, it should be noted that the rheology (Newtonian or non-Newtonian), the flow regime (fully turbulent or not) and the geometrical similarity of the systems involved need to be checked and compared. In addition, the high power cost of agitation and/or aeration in maintaining a given $k_L \cdot a$ value may not be justified by the resulting increase in product yield, and in some cases a product may be biosynthesized at its maximum rate even when the oxygen demand of the organism is not satisfied.

Other criteria of possible relevance are impeller tip speed (section 4.5), for shear-sensitive organisms and very viscous cultures; mixing time (section 4.4); turbulence or Reynolds number (section 4.4); and power input per unit volume (section 4.4).

4.6.3 Interaction of criteria

The effects of maintaining one criterial parameter constant on other environmental parameters can be seen by considering some of the correlations cited in earlier sections.

Scale-up on the basis of maintaining a constant power input per unit volume was used for many years as the rule-of-thumb '1 horse-power per 100 gallons'—about 1.7 kW m^{-3}. This worked because on fermenters with single impellers, the exponent of power per unit volume (x in eqn 4.36) is only slightly less than one, so that the rule was virtually equivalent to using $k_L \cdot a$ as the criterion, for a constant superficial gas velocity. With the advent of larger fermenters with multiple impellers, exponents of about $\frac{1}{2}$ to $\frac{2}{3}$ are found more appropriate. As discussed in section 4.2.4, the lowest impeller of a multiple impeller stack has a shearing function to effect gas dispersion and the upper impellers a mixing function, so that they contribute to the power dissipation, but less to gas dissolution.

For geometrically similar vessels, the vessel volume V is proportional to D_i^3. With constant impeller tip speed, $N \cdot D_i$, the impeller speed is proportional to $V^{-1/3}$. With constant power input per unit volume, for fully turbulent systems the impeller tip speed increases with vessel size, since, from eqn 4.20, N is proportional to $V^{-2/9}$. For very viscous cultures, eqn 4.41 shows that a constant OTR is maintained by maintaining impeller tip speed to a constant exponent 1.6. As was shown in section 4.4.1, maintaining constant mixing time t_m leads to the power input per unit volume being increased roughly in proportion to vessel volume, whereas maintaining constant $k_L \cdot a$ is found in practice to require decreasing power input per unit volume as vessel size increases. For constant Reynolds number, Re, in fully turbulent conditions, from eqns 4.16 and 4.20, with increasing scale:

(i) Power input per unit volume decreases, and is proportional to $V^{-4/3}$
(ii) Impeller tip speed, $N \cdot D_i$, is inversely proportional to the impeller diameter and so decreases
(iii) The impeller speed is inversely proportional to the square of the impeller diameter.

As pointed out in section 4.2.5, the superficial gas velocity is scale-dependent if a constant value of volume gas flow-rate per unit culture volume is maintained, as it is then proportional to $V^{1/3}$, so that the superficial gas velocity is closer to the limiting value in large fermenters than in small ones of a similar design.

4.7 Fermentation processes in biotechnology

Examples of the applications of the principles discussed in the foregoing sections can be found in the range of commercially important industrial biological processes discussed in Chapter 2. Features of some of these processes will be used here to illustrate points of fermentation engineering principles.

4.7.1 *Brewing* (see 2.4.1.2 for fermenting organisms)

The brewing industry provides examples of improved fermenter design evolving from an understanding of the details of process mechanisms, and of the choice of operating mode.

Although a small but significant proportion of beer brewed in the UK is produced by a continuous process, batch fermentation has been selected for most of the recent brewery capital investment programmes. The flirtation with continuous brewing in the late 1960s led to intensive investigation of the operational problems of the traditional batch process, with the result that the new developments were in the direction of accelerated batch processing. The acceleration resulted from improved equipment design, coordination of batch sizes to avoid production bottlenecks and separation of multiple process operations into individual units designed specifically for each process, rather than as an overall compromise. In addition, the brewing industry operates hygienically rather than aseptically, and the added infection hazards resulting from continuous operation led to the abandonment of some continuous processing plant. One particular design, the plug-flow continuous 'tower' fermenter, depended on the ability of a particular strain of yeast to flocculate, and loss of this ability during a fermentation led to loss of the culture. The overriding advantage of batch operation, however, is its flexibility. The demand for beer is seasonal, requiring a variable production rate, and in addition, most breweries produce a wide range of products, with a few consistently-selling lines and a plethora of minor speciality products. Such a production pattern does not allow full benefit to be derived from the advantages of continuous operation, even though the inhibitory effect of alcohol on the fermentation rate appears advantageous at first sight.

The concept of accelerated batch processing led to the development of several new types of equipment, among them the so-called 'conical' fermenter. The use of a tall, relatively narrow vessel rather than the traditional flat tank conferred several advantages. The longer path through which the bubbles of carbon dioxide evolved by the fermentation rise through the liquid creates greater agitation, leading to a greater degree of homogeneity and the closer control of temperature. The fermenter has a clean, uncluttered interior, as the agitation enables temperature control to be maintained with external attemperator jackets only, allowing easier in-place cleaning. The actual diameter chosen is based on the maximum allowable width of load to be transported by road.

4.7.2 *Penicillin manufacture* (see also 2.5.1 for microbiological aspects)
The 'clean' fermenter interior is preferred for the mould fermentation in penicillin production, so that external cooling jackets are used for temperature control. The lower heat-transfer coefficient and heat-transfer area obtained with jackets as opposed to internal coils is compensated by using chilled water or alcohol–water, so that the increased temperature difference gives the required heat-transfer rate in accordance with eqn 4.4. Internal coils are considered to interfere with vessel cleaning and with air distribution in the mechanically-agitated porridge-like mould culture. The liquid acquires this consistency because the turbulence from agitation tends to roll up the mycelium into small pellets, which allows the broth to be separated from the mycelium by the relatively cheap and simple operation of precoat filtration before solvent extraction of the penicillin. The mode of

fermenter operation is serial or incremental culture, whereby the penicillin-productive phase is prolonged by controlled addition of supplementary growth medium. The rate of addition is such as to maintain the nutrient level sufficient for penicillin synthesis, but not high enough to encourage massive mycelial growth. Prolonging the fermentation excessively can, however, result in some autolysis of the mycelium, which can cause problems in later separation stages. Batch-based operation is used because of the growth phase in which the mould has greatest penicillin productivity, and because in ethical pharmaceutical manufacture, identification of product components is essential.

4.7.3 *Biological waste-water treatment*

Biological waste-water treatment is the area of biotechnology which is most widespread and involves the greatest capital investment, but its objectives and process conditions are in several respects the opposite of the other processes considered in this section. The overall aim of the process is the removal and breakdown of material, rather than its production, and is based on the apparently straightforward process in which a mixed microbial population utilizes polluting substances as nutrients. Biological waste-water treatment provides a classic example of the development of successful, large-scale processes in a vital area of biotechnology from the coordinated application of engineering and microbiology. Waste-waters containing polluting substances are brought into contact with a dense population of suitable microorganisms for a time sufficient for the organisms to break down and otherwise remove pollutants to the required extent.

The process is run non-aseptically, generally in open concrete tanks, and a wide range of different organisms is present which tends to be self-selecting. The microbial population is largely moribund, giving it a tendency to flocculate. Floc formation enables the organisms to be separated from the treated waste-water cheaply, rapidly and efficiently, by settling, and also provides an additional pollutant removal mechanism. Suspended, colloidal and some dissolved material is removed from the waste-water by adsorption and agglomeration on to the microbial flocs, and this material and other dissolved nutrients are broken down by microbial metabolism. Part of this material is converted to simple substances, such as carbon dioxide, and part to new biomass, a proportion of which is then broken down by endogenous respiration. The nutrient concentration, and consequently the oxygen demand, of waste-waters is generally an order of magnitude or more lower than in fermentation growth media and closer to the starvation level of the organisms. The problem of satisfying the oxygen demand of the process is not so much one of supplying sufficient oxygen, but of supplying it cheaply. Overall, the processes are designed to deal with large volumes of dilute liquids, separating them into purified water and a 'pollution concentrate' in the form of biomass.

4.7.4 *Single-cell protein (SCP) production* (for microbiological and chemical aspects see 5.2.5 and 2.4.6.10)

The ICI Ltd 'Pruteen' process produces SCP in the form of bacterial biomass for animal-feed formations. The process came on stream in 1981 and is a recent example of advanced fermentation engineering. It is particularly interesting in that it is a high-volume, intensive, relatively low-value-product process, like most of the chemical industry, where the high productivity required exacerbates the problems of asepticity in fermentation. The relatively low-value product is used in a highly competitive field, so that maximum productivity is essential and is provided only by large-scale continuous fermentation. This involves the institution and monitoring of strict aseptic procedures as outlined in section 4.2, as it has been found that infecting organisms can stabilize and form a large proportion of the biomass (Smith, 1980). This is unacceptable, since the desired organism was developed for its efficient utilization of the methanol substrate, its ability to flocculate and so simplify biomass separation, and its lack of harmful effects in extensive animal feeding trials, so that infection by an unknown organism could impair these essential biomass characteristics. The spent liquor after biomass separation is recirculated to the fermenter following supplementation with fresh methanol and nutrient salts.

The process uses a $1500 \, m^3$ air-lift fermenter, of diameter 7 m and overall height 60 m (Figure 4.9). The liquid depth of 45 m gives a high hydrostatic pressure at the bottom, which enhances dissolution of air injected through a sparger at the bottom of a draught tube 6.6 m in diameter. The pressure difference between the aerated liquid in the draught tube and the bubble-free liquid in the annular downcomer drives the aerated liquid upwards, past cooling tubes and through a series of perforated baffle-plates which induce turbulent fluid flow. The turbulence enhances gas–liquid contact, for air dissolution and carbon dioxide stripping, and rapid dispersion of the methanol substrate, which is injected at intervals up the column at rates controlled in response to dissolved oxygen measurements. The liquid passes into the broader (11 m diameter) top section of the column for disentrainment of the exhaust gases, and the bubble-free liquid recirculates down the annular downcomer. As liquid passes through the vessel it circulates several times in the above manner, undergoing successive high- and low-pressure cycles, so that the unit is sometimes called the 'pressure-cycle fermenter'. An adaptation of this principle has been developed for use in waste-water treatment, called the 'Deep-shaft' process, where the enhanced air dissolution rates enable very short contact times to be used.

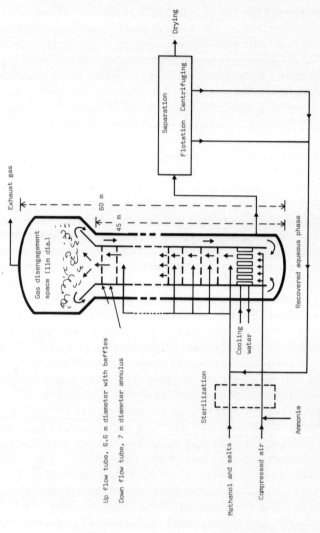

Figure 4.9 ICI Ltd 'pressure cycle' airlift fermenter for bacterial SCP from methanol.

The culture passes from the downcomer to a separation stage where the flocculent biomass is separated by flotation and centrifuging before drying, the recovered liquid being recycled to the fermentation. The handling of the wet solids between centrifuging and drying is the most problematical stage of the process in terms of scale-up and prediction of plant performance (Andrew, 1981). The air-lift fermenter was chosen because, with a power input of $1.6\,\mathrm{kW\,m^{-3}}$ totalling 2.5 MW, mechanical agitation would have required multiple shafts to transmit this amount of power, with consequent multiple infection hazards. The fermenter vessel is of mild steel, with the shell thickness varying from 30 mm at the base to 16 mm at the top of the column, and additional circumferential stiffeners are welded on the outside of the shell to enable the vessel to withstand the vacuum produced during post-sterilization cooling. This very large vessel acts as a large heat sink during sterilization, and at first, incomplete sterilization was a major cause of fermenter downtime. Control of pH in the process is effected by injection of ammonia gas into the input air stream prior to sterilization.

4.7.5 *Outlook for biotechnology*
The processes considered in the previous sections cover ranges of material value and organism selectivity—high-value to low-value products, high-value, low-value and negative-value raw materials, and strictly aseptic, hygienic and completely non-aseptic operation. They have a key feature in common, however—they all involve dilute solutions. After years of development, the penicillin process produces a 2% solution of the high-value product, and the most recent, the bacterial SCP process, gives an output containing about 3–4% of a low-value product. In chemical processes, such low concentrations are usually found in reject streams. Dilute reaction media involve large reactors, with consequent high capital costs, high pumping power costs for moving large volumes of liquid and high product separation costs.

The key area of activity of biotechnology in the foreseeable future is thus likely to remain the manufacture of high-value, low-volume products, such as pharmaceuticals, rather than of low-value, high-volume products such as organic solvents and feedstocks and fuel supplements.

4.8 Summary

Fermentation engineering is the branch of biotechnology dealing with the design, development, construction and operation of plant and equipment used in industrial biological process. Its principal task is to provide a growth environment in large-scale operation equally favourable to that in the original laboratory-scale process and at minimal cost. Large-scale processing involves problems of bulk handling and scale-up, and the engineer applies the principle of Occam's Razor to meet the ultimate criterion of lowest unit cost of product. The principle of aseptic plant design and the relative merits of batch and continuous processing must be assessed, and the general engineering principles of heat transfer, in temperature control and heat sterilization, mixing, mass-transfer and oxygen supply, and scale-up taken into account with particular reference to stirred-tank and gas-sparged fermenters, with illustrations from four different, representative large-scale biological processes.

References

Anderson, J. G. and Blain, J. A. (1980) Novel developments in microbial film reactors. In *Fungal Biotechnology*, eds. J. E. Smith, D. R. Berry and B. Kristiansen, Academic Press, London, 125–152.
Andrew, S. P. S. (1981) The engineering of the ICI 'Pruteen' process. *Proc. Conf. Future Developments in Process Plant Technology*, I.Mech.E., C3/81, 17–21.
Atkinson, B. and Lewis, P. J. S. (1980) The development of immobilised fungal particles and their use in fluidised bed fermenters. In *Fungal Biotechnology*, eds. J. E. Smith, D. R. Berry and B. Kristiansen, Academic Press, London, 153–174.
Banks, G. (1977) Aeration of mould and streptomycete culture fluids. In *Topics in Enzyme and Fermentation Biotechnology*, ed. A. Wiseman, Ellis Horwood, Chichester, vol. 1, 72–110.
Banks, G. (1979) Scale-up of fermentation processes. *Ibid.*, **3**, 170–266.
Blakebrough, N. (1967) Industrial fermentations. In *Biochemical and Biological Engineering Science*, ed. N. Blakebrough, Academic Press, London, **1**, 25–48.
Burrows, S. (1979) Baker's yeast. In *Microbial Biomass*, ed. A. H. Rose, Academic Press, London, 32–64.
Calderbank, P. H. and Moo-Young, M. B. (1959) The prediction of power consumption in the agitation of non-Newtonian fluids. *Trans. I. Chem. E.* **37**, 26–33.
Evans, L. (1974) *Selecting Engineering Materials for Chemical and Process Plant*. Business Books, London.
Finn, R. K. (1967) Agitation and aeration. In *Biochemical and Biological Engineering Science*, ed. N. Blakebrough, Academic Press, London, **1**, 69–101.
Kristiansen, B. and Bu'Lock, J. D. (1980) Developments in industrial fungal biotechnology. In *Fungal Biotechnology*, eds. J. E. Smith, D. R. Berry and B. Kristiansen, Academic Press, London, 203–224.
LeGrys, G. A. and Solomons, G. L. (1977) Patent appl. 23128.
Metzner, A. B. and Otto, R. E. (1957) *AIChEJ* **3**, 3–10.
Michel, B. J. and Miller, S. A. (1962) *ibid.* **8**, 262–266.
Norwood, K. W. and Metzner, A. B. (1960) *ibid.* **6**, 432.
Richards, J. W. (1968) *Introduction to Industrial Sterilisation*, Academic Press, London.
Schügerl, K., Lücke, J. and Oels, U. (1977) Bubble column bioreactors. In *Adv. Biochem. Eng.*, eds. T. K. Ghose, A. Fiechter and N. Blakebrough, Springer, Berlin, **7**, 1–84.
Schügerl, K., Lücke, J., Lehmann, J. and Wagner, F. (1978) Application of tower bioreactors in cell mass production. *Ibid.* **8**, 63–132.
Smith, S. R. L. (1980) Single cell protein. *Phil. Trans. R. Soc. London, Ser. B*, **290**, 341–354.
Solomons, G. L. (1980) Fermenter design and fungal growth. In *Fungal Biotechnology*, eds. J. E. Smith, D. R. Berry and B. Kristiansen, Academic Press, London, 55–80.
Solomons, G. L. (1969) *Materials and Methods in Fermentation*, Academic Press, London.
Steel, R. and Maxon, W. D. (1962) *Biotech. Bioeng.* **4**, 231–240.
Thomas, A. and Winkler, M. A. (1977) Foam separation of biological materials. In *Topics in*

Enzyme and Fermentation Biotechnology, ed. A. Wiseman, Ellis Horwood, Chichester, vol. 1, 43–71.

Wilkinson, W. L. (1960) *Non-Newtonian Fluids*. Pergamon, Oxford.

Winkler, M. A. (1981). *Biological Treatment of Waste-Water*. Ellis Horwood, Chichester.

Winkler, M. A. (1983) Limitations of fermentation processes for utilisation of food wastes. In *Topics in Enzyme and Fermentation Biotechnology*, ed. A. Wiseman, Ellis Horwood, Chichester, vol. 7, 271–306.

Winkler, M. A. (1988) Optimisation and time-profiling in fermentation processes. In *Computers in Fermentation Technology*, ed. M. E. Bushell, Progr. Ind. Microbiol. **25**, Elsevier, Amsterdam, 64–123.

5 Application of the principles of enzymology to biotechnology
ALAN WISEMAN

5.1 Features of enzymes in relation to biotechnology

5.1.1 Introduction

Enzymes are the biological catalysts used by living cells to achieve a variety of chemical conversions recognized as the 'chemistry of life'. Every cell contains a large number of enzymes, each with an ability limited to the conversion of a particular chemical molecule or portion of molecule to a modified or cleaved version of that molecule. In chemical terms many of these reactions are simple ones, for example the addition of hydrogen atoms, or the addition of water. The value of biochemical routes lies in the sequential conversion in a metabolic pathway of a starting material into the required product. The intermediary metabolism along this route is in some cases of relatively minor interest to the biotechnologist, who is seeking the most economical harnessing of nature's catalysts to a practical problem.

5.1.2 Advantages of using enzymes for manufacture of products

Enzymes carry out a rapid conversion of their substrate at moderate temperatures and near-neutral pH, usually with high specificity both in selection of substrate from a complex mixture, and in the chemical conversion actually effected. Foodstuffs are often the substrates, as for the well-known enzymes produced in the stomach or intestine of animals. Hydrolysis of carbohydrates, proteins and fats respectively are easily achieved there by particular enzymes, and similar enzymes (usually obtained from microorganisms) are used in many such reactions in the food industry.

High specificity requires the use of a single enzyme, however. Therefore extensive purification, with subsequent increase in cost, is necessary, if such a single enzyme activity has to be achieved by removal of contaminating enzymes from the crude mixture isolated from the many microbial, plant or animal sources of enzymes. Prior purification of crude extracts of plants or microbial culture media is in some cases now a legal requirement. This might be so where addition of the enzyme to a human food necessitated the guaranteed absence of toxic components or contaminants. For example,

Table 5.1 Some enzymes used in industry (see also Chapter 7)

Enzyme	Use
Bacterial glucose isomerase	Glucose→invert sugar (i.e. fructose formation)
Bacterial α-amylase } Fungal amyloglucosidase	Starch→glucose
Fungal α-amylase	Partial degradation of starch in supplementation of amylase-deficient flour for breadmaking
Microbial rennets	k-Casein→para-casein (in milk curdling for cheese manufacture)
Bacterial proteases	Removal of protein-based stains and laundering (in biological washing powders)
Papain (from papaya melon)	Several protease applications including meat tenderization and dehazing of beer
Cellulases	Cellulose→glucose
Fungal pectinase	Pectin degradation (in fruit and vegetable processing)
Aminoacylase (immobilized)	Resolution of DL-amino acids to produce L-amino acids for food supplementation
Glucose isomerase (immobilized)	Production of invert sugar and of high fructose syrups from glucose
Penicillin acylase (immobilized)	Hydrolysis of penicillin-G to make 6-aminopenicillanic acid for production of new penicillins

some fungi can under some conditions of growth produce mycotoxins which are highly poisonous or carcinogenic; the best-known case is that of aflatoxin production by *Aspergillus flavus*. Highly sensitive detection and assay procedures are now available for such toxins, based for example on immunoassay techniques (these use antibodies, produced in animals or animal cell cultures, which recognize the toxin and bind tightly to it). Some enzymes used in industry are listed in Table 5.1.

Another important feature and advantage of using enzymes is that their action can be readily controlled, not only by careful setting of reaction conditions, but also by the ease of stopping the reaction by heating, where and when appropriate. This heating stage destroys the enzyme, however. Single use of an expensive purified enzyme would make any large-scale industrial process uneconomic. Re-use of the enzyme is possible if it can be recovered in active form at the end of the required processing step. Such a re-use facility can be achieved by *immobilization* of enzyme to a solid support (see Chapter 7). Immobilization of enzyme in favourable circumstances can lead to the use of a chemical engineering 'unit operation' approach to a process involving enzymes. Furthermore, stabilization of the enzyme is often in practice achieved by this prior step.

Enzymes, especially immobilized enzymes, are therefore likely to be chosen for the following reasons.

5.1.2.1 *Process improvement and speed-up.* Many aspects of a biotechnology-based process can be greatly improved, using the appropriate enzyme.

Simple hydrolases, the enzymes that add water to the linking bonds in polymers, will break down these macromolecules, eventually to their constituent monomers in some cases. Thus amylases will degrade starch, and proteases will degrade proteins. Such degradation, involving a lowering of chain length (and molecular weight) of the polymeric chain, will change many of its features, especially those associated with its polymeric and colloidal properties. Increase in solubility in water and decrease in viscosity is the change most likely to be noticed in a process, and this may remove problems of pumping, filtration and effluent disposal, which may be major barriers to an otherwise convenient and economically-viable process.

5.1.2.2 *Product improvement.* Requirements for a saleable product depend upon many factors including consumer acceptability and government safety requirements for avoiding toxicity. These will be different from country to country, so that considerable flexibility of approach is essential.

An enzymic step may be useful in changing the taste or texture of a foodstuff, or its colour or aroma. It is the composition of the foodstuff that is being altered for these purposes, and other useful changes may be achieved by the manufacturer for a variety of reasons connected with nutritional and digestibility aspects of the product, or, for example, its shell-life.

5.1.3 *Choice and control of enzymes in applications*

Even an impure enzyme will select its substrate from a complex mixture of materials, bind to it and carry out the conversion characteristic of that enzyme. The choice of a suitable enzyme for a particular application will be determined by the conversion required and the extent of our tolerance of side reactions caused by other enzymes present. Purification to eliminate these side reactions is costly, and in addition many enzymes (although not hydrolases) require cofactors such as NAD^+ (nicotinamide adenine dinucleotide) which are themselves very expensive to buy. A good source of the identified required enzyme is essential. Usually the required enzyme is a well-known and characterized commercially available enzyme. If it needs to be isolated, its distribution in animal, plants or microorganisms is recorded in the extensive scientific literature of biochemistry, microbiology and biotechnology (see Chapter 6).

Unlike other substances, enzymes are bought and sold on the basis of activity, not weight *per se*. If the activity is unstable and has been lost in storage or transit, the enzyme is valueless. The study of protein conformation in relation to denaturation of enzymes is vitally important therefore to enzyme biotechnology. Thus, for example, each enzyme has an optimum pH, usually in the range pH 5–9, at which its rate of reaction is the fastest: enzymic stability in storage or in actual use, however, may not be the most prolonged at this optimum pH. Similarly each enzyme has an 'optimum' temperature. Reaction rates increase with rising temperature (approximately doubling for each 10°C rise), but concurrent denaturation of enzyme occurs at higher temperatures. It is the balance between these two opposing effects that is characteristically achieved by each enzyme at a particular 'optimum' temperature. Another characteristic of a particular enzyme is its maximum velocity (abbreviated to V_{max} or V) achieved in the presence of excess substrate (saturation achieved of enzyme active sites) at any particular temperature and pH chosen to study the enzyme. The Michaelis constant (K_m) is defined as the substrate concentration required to reach half the maximum velocity (K_m values are usually in the range 10^{-2} M–10^{-6} M, and high affinity between substrate and enzyme is usually reflected in a low value, e.g. 10^{-5} M or 10^{-6} M).

It is most important to appreciate that the industrial application of enzymes is carried out in a complex situation. Here the classical kinetic parameters and considerations outlined above may be of little *practical* significance because of the presence of a variety of inhibitors, and other complications. The practical kinetics of the system must therefore be determined *under working conditions*!

5.2 Applications of enzymes in biotechnology

5.2.1 *Large-scale industrial applications*

Enzymes in soluble form have been used in the food industry for many years. This is especially so in the baking and brewing industries, the latter being the best example of traditional biotechnology, prior to recent developments and interest in biotechnology as a novel approach to production of materials from renewable energy sources such as plants and microorganisms.

5.2.1.1 *Uses of amylases.* The ability of α-amylases to cause mid-chain random degradation of starch has been of vital importance to several industries, including baking, brewing, distilling, textile and paper manufacturing.

In the baking industry, the production of carbon dioxide by the yeast requires the fermentative breakdown of glucose (the other product is ethanol, most of which evaporates off, but which provides the flavour of newly-baked bread). Extra quantities of small molecule sugars, including glucose, are made available to the yeast by the addition of α-amylase produced from fungi (*Aspergillus* species) to the wheat flour. This is especially useful if the wheat is found to be deficient in the α-amylase normally produced in the wheat during germination. Supplementation of flour with fungal α-amylase is particularly useful in the USA in dry years when inadequate germination of the wheat occurs to provide adequate levels of α-amylase in relation to the baking of bread. Many important advantages are claimed for the amylase-supplemented bread flour, including improvements in bread colour, texture, and shell-life. The action of the fungal α-amylase is rapidly destroyed during the baking processes, but some amylases, e.g. those from bacteria, are much too heat-stable for this application.

Bacterial α-amylases are used extensively in the other applications of amylases, however. Starch is much more readily dissolved in hot water following a prior gelatinization step using thermostable bacterial α-amylase at 100°C (or higher in some cases). Particularly thermostable enzymes are readily obtainable from thermophilic bacteria (which can live at relatively high temperatures). Further degradation of starch occurs in the presence of the amylase, and growth media rich in glucose and oligosaccharides are produced. Extra sugars are produced in this way from adjuncts of wheat flour added to supplement brewing growth media produced from barley malt. In these industries, extra amino acids and peptides are produced by the use of proteases (see below), which also supplement the yeast growth media.

The applications of bacterial amylases in the textile and paper manufacturing industries depend on the usefulness of gluey degradation products (dextrins) produced from starch. Starch sizes (coats) are used in the textile industries to preserve the twist of the fibre and to add weight during weaving. These starch sizes made from amylase-degraded starch also need to be removed later, by further amylase action. Similarly, in the paper industry, starch is pretreated to give a lowered-viscosity product needed for sizing and coating processes: these sizes are removed later by further treatment with bacterial α-amylases.

5.2.1.2 *Uses of proteases.* An enormous variety of applications exist for these protein-degrading enzymes. Enzyme washing powders contain a small amount of bacterial protease (from *Bacillus subtilis*) to provide the biological catalytic action needed to dissolve dried bloodstains, for example, which are very difficult to remove by any other acceptable means. The same enzyme (subtilisin) is used in the leather industry to loosen hair from the hide, or in the textile industry to recover wool from sheepskin.

Papain is a widely used protease prepared from the latex of the fruit of the tropical fruit tree *Carica papaya* (Brocklehurst *et al.*, 1981). It can be used to tenderize meat by preferential degradation of tough connective tissue rather than the actinomyosin component of meat. Canned meat or steak may be tenderized in this way, and the action of the added papain continues for a significant time while the meat is cooking before this relatively thermostable enzyme is destroyed. Another application of papain is in the chillproofing of beer. Precipitation of a protein-tannin complex is likely to occur if beer is stored cold. Prior degradation of this protein component, which may be derived for example from the barley malt used in brewing, prevents the 'chill haze' problem which the consumer finds unattractive.

Several useful proteases have been obtained from animal sources. These include pepsin, trypsin and chymotrypsin, used as digestive aids and for predigestion of baby foods. Rennin, derived from the fourth stomach of the unweaned calf, breaks a single peptide bond in the casein component of milk protein, causing it to co-precipitate with calcium ions. This is the basis of cheesemaking, although shortage of rennin led to an extensive search for a suitable replacement, and this technically very exacting enzymic requirement has now been developed by use of microbially-derived rennin substitutes, some made by cloning the appropriate calf gene into microorganisms. This is an example of genetic engineering to produce a useful product. Rennin itself may still be used, however, in the home for the 'clotting' of milk to form junkets.

5.2.1.3 *Uses of enzymes in sugar industries.* Very large amounts of glucose are produced by the degradation of starch from a variety of sources on a worldwide basis. Fungal amylases are used that have an amyloglucosidase (glucamylase) content adequate to degrade the α-1,6 cross links in the starch molecule. α-Amylases break only the α-1,4 main chain links in the starch, so that it is not completely degraded to glucose—the product also contains limit dextrins, which amyloglucosidase activity is able to degrade. Complete conversion of starch to glucose is an important industry, but as glucose is not very sweet, the worldwide demand for sweetness has led to another process, the conversion of glucose to fructose (very sweet) using glucose isomerase (see 7.2). Normally this enzyme proceeds to the position of equilibrium of the isomerization, so that a roughly equal mixture of glucose and fructose is obtained. This mixture, known as invert sugar, has been produced for many years by the acid hydrolysis of sucrose and could also be achieved by the enzymic splitting sucrose using yeast invertase.

Much recent academic and industrial study has centred around the properties and use, in soluble and immobilized form, of glucose isomerases produced by bacteria. Millions of tons of invert sugar and the related high fructose syrups are produced by the use of this enzyme, especially in the USA.

Another important enzyme, which attacks the disaccharide lactose, is β-galactosidase (lactase, produced by some yeasts). Lactose is hydrolysed to its constituent monosaccharides, glucose and galactose, by lactase. Free or immobilized forms of this enzyme can be used for example to remove lactose from milk, because it can cause intestinal discomfort in the populations of some countries.

Glucose itself can be removed from some foodstuffs, for example prior to drying where glucose would cause discoloration, by fungal glucose oxidase. This enzyme uses glucose and oxygen, so that either substance can be removed entirely in the presence of an excess of the other. Complete removal of oxygen can be achieved in the presence of excess glucose, and this is desirable in many cases to prevent autoxidation (rancidity) problems in foodstuffs. The same enzyme is used to detect and assay glucose, and is incorporated into quick test procedures for glucose in blood and urine (diabetes test).

The reader is referred to the more detailed accounts of enzyme isolation and purification (Chapter 6) and to the enzyme engineering aspects of the application of immobilized enzymes and cells (Chapter 7). A first introduction to immobilized enzymes is available in Trevan (1980).

References

Brocklehurst, K., Baines, B. S. and Kierstan, M. P. J. (1981) In *Topics in Enzyme and Fermentation Biotechnology*, ed. A. Wiseman, vol. 5, Ellis Horwood, 262–335. (Vols. 1–10, 1977–85, contain many other reviews—a more general review volume is *Handbook of Enzyme Biotechnology*, ed. A. Wiseman, 2nd edn., Ellis Horwood, Chichester, 1985).

Trevan, M. D. (1980) *Immobilized Enzymes*. John Wiley, New York.

6 The biotechnology of enzyme isolation and purification
C. BUCKE

6.1 Introduction

Enzymes from most types of organism find commercial uses for which, in most cases, they must be separated from the cells which produced them and purified. In all cases the enzyme will be purified as little as is necessary to remove interfering activities, but for some tasks, for example particular analytical uses, high degrees of purity are required. About 1500 tonnes of enzymes are sold worldwide per annum. Four enzymes comprise about 1250 tonnes of these—bacterial protease, used in detergents, and glucoamylase, bacterial α-amylase and glucose isomerase, all used in the manufacture of glucose and fructose syrups from starch. Hundreds of others, highly purified and therefore expensive, find worldwide sales of only milligrams per annum for sophisticated research purposes.

Enzyme isolation and purification are processes involving a multiplicity of techniques; this chapter describes the techniques and their application (see Lilly, 1979; Wang et al., 1979).

6.2 Enzyme sources

Enzymes are present in all living things and, if sufficient care is taken to protect them, they can be isolated and purified from any organism (see Darbyshire, 1981). The great bulk of enzymes used in industry are microbial in origin, but there are exceptions, such as the various plant proteases (including papain, bromelain and ficin) and the animal proteases such as rennin and pepsin. Enzymes for research purposes are isolated from all types of organism, some of which are less easy to obtain, for example sulphatase from the digestive juice of snails, and various enzymes from snake venoms.

6.3 Release of enzymes from cells

Many microbial enzymes that degrade polymeric substrates are found extracellularly, and therefore separation of those enzymes from cellular material

Table 6.1 Factors likely to inactivate enzymes during isolation and purification.

Inactivating factor	Enzyme source	Frequency of occurrence	Counteracted by
Heat	Any	Almost universal	Cold
Cold	Any	Rare	Warmth
Protease action	Most	Common	Protease inhibitors Cold Rapid separation
Oxidation products of phenolics	Plants, fungi	Fairly common	Reducing agents
Oxidation	Any	Common	Reducing agents Avoidance of stirring
Shear	Any	Fairly common	Avoidance of shearing
Protein dilution	Any	Fairly common	Rapid concentration
Loss of stabilizing factor(s) e.g. metal ion, substrate	Any	Fairly common	Inclusion of that factor in media
Indigenous, specific inhibitors	Plants, bacteria	Rare	Separation from inhibitor
Heavy metals	Any	Rare	Chelating agents Good practice
Phase change, e.g. foam	Any	Common	Minimal agitation

can be achieved simply by filtration or centrifugation. Before purification of intracellular enzymes, it is necessary to release them into solution from the cells that produced them. Before dealing with means of releasing enzymes from cells, it is important to note the hazards which may befall enzymes during the purification process, in particular in the early stages and which may cause very significant loss of activity (Table 6.1). In general, factors must be avoided that disrupt the conformation of the enzyme protein, allowing the protein to 'unwind', thus increasing its vulnerability to destruction by chemical means or by protease attack (see Tombs, 1985; Ahern and Klibanov, 1985).

6.3.1 Sources

6.3.1.1 *Plants.* In general, higher plant tissues are not satisfactory materials from which to attempt to isolate enzymes. Plants have no means of excreting waste materials and accumulate these in vacuoles. On disruption of cells (which may require the input of large amounts of energy to smash tough cell walls), the vacuole contents are released and come into contact with enzymes, with undesirable effects. For example oxidation products of phenolics are able to bind to enzyme proteins and in some cases remove all activity: this effect varies greatly from enzyme to enzyme. Such inactivation may be partially overcome by including reducing agents such as β-mercaptoethanol, ascorbate or thioglycollate in the extraction medium. Young plants tend to have a lower phenolic content, and some species, in particular spinach, have very low phenol oxidase activities. Consequently, spinach leaf is the generally preferred starting material for the purification of photosynthetic carbon reduction cycle enzymes.

Where enzyme inactivation results from the action of other enzymes it is obviously an advantage to keep the temperature as low as possible during extraction.

6.3.1.2 *Animal tissues.* The primary hazard to animal cell enzymes is hydrolysis by proteases released by the necessary disruption of the cells. Their effects may be minimized by inclusion of additional proteins, such as albumins, in extraction media (although these may have to be removed later in the purification process), by keeping the tissue cool during extraction, and by isolating the enzyme required as rapidly as possible.

6.3.1.3 *Microbial cells.* The vast diversity of microbes makes it even less easy to generalize about the precautions which should be taken during the extraction of their enzymes—destruction by proteases is always possible, and many fungi are rich in phenol oxidases. Thus, rapidity of operation and low temperatures during isolation are advantageous. It should not be forgotten, however, that a few enzymes are cold-labile!

6.3.2 *Extraction by physical methods*

6.3.2.1 *Homogenization.* In general, the tissues of higher organisms, mammals, higher plants and multicellular algae are tough, but can be disrupted satisfactorily in quantity using comparatively violent means. Familiar domestic mincers and liquidizers are suitable for this task, but often are not sufficiently robust for continuous use. Waring blenders are liquidizers of appropriate strength. For large-scale extraction of animal and plant tissues, hammer mills, suitably modified, are commonly used.

Once disrupted, animal and plant cells release enzymes readily, although in some cases inclusion of detergent to release membrane-bound enzymes is advantageous. The best results are obtained using the softest tissues of the youngest organisms in order to minimize the time required for comminution and thus reduce the introduction of oxygen, heating of the sample and the time of exposure during which degradative enzyme activity may occur.

The equipment used for plant and animal tissues is not usually satisfactory for the disruption of microbial tissues which have greatly tougher cell walls (see below).

6.3.2.2 *Agitation with abrasives.* Ball-milling provides an excellent means of disintegrating cells. Vessels containing glass beads vibrating rapidly provide shear forces because of velocity gradients and by collisions between beads and organisms. The extent of disintegration depends on the rate of stirring or vibration, the ratio of beads to cells, the nature of the cells, the size of the beads and the contact time. This principle can be used on a very small scale (e.g. in disrupting organisms in conditions of strict containment), in continuous equipment capable of processing kilogram quantities of cell pastes per hour and on intermediate scales. Equipment most frequently encountered includes the Mickle Shaker (two small vessels vibrated rapidly); the Dyno-Mill, a much larger continuous or batch apparatus in which beads are agitated by rapidly rotating discs; and the Bead Beater, an intermediate-sized vessel in which beads are stirred rapidly. The principal hazard to enzymes in this type of apparatus is heat; overheating of small-scale equipment may be limited by working in a cold room, but Dyno-Mills require powerful refrigerators to maintain low temperatures. Such agitation is capable of disrupting the toughest of cells, given time. Solid shear methods are particularly suitable for disrupting filamentous organisms.

In the case of the disruption of yeast cells, with plug-flow through the equipment and with batch operation, first-order kinetics are observed; the rate of protein release is directly proportional to the amount of protein that is unreleased:

$$\frac{dR}{dt} = k(R_m - R)$$

where R is the weight of protein released per unit weight of packed yeast, and R_m is the maximum measured protein release.

6.3.2.3 *Liquid shear.* On forcing a cell suspension at high pressure through a narrow orifice, the rapid pressure-drop provides a very powerful means of disrupting the cells. In practice it is simple to design equipment to subject the cell suspension to shear forces before releasing the pressure. By 'tuning' the equipment, cells may be disrupted completely or only sufficiently to release periplasmic enzymes. The most commonly-encountered liquid-shearing equipment is the Manton–Gaulin APV homogenizer, a positive displacement piston pump which draws the cell suspension through a valve into the pump cylinder and forces it through an adjustable discharge valve with a restricted orifice. In the case of yeast cells, Hetherington *et al.* (1971) demonstrated that the release of protein is described by a first-order rate equation:

$$\log \frac{R_m}{R_m - R} = KNP^{2.9}$$

where R_m is the maximum amount of soluble protein obtainable, R is the amount of soluble protein released by N passes through the homogenizer, K is a temperature-dependent rate constant, and P is operating pressure.

Clearly, the approximately cube-power relationship between pressure and rate of disruption indicates the requirement for high-pressure units for efficient cell disintegration. A single pass is always to be desired, as the further comminution of cell debris may cause difficulties with clarification during further processing.

Once again, heating is the principal hazard to enzymes in this type of equipment. This is minimized by cooling the cell suspension as far as possible. It is most important that frozen lumps or foreign bodies are eliminated, or equipment blockage will result. Consequently, liquid shear devices are more suitable for treating unicellular organisms than mycelial organisms, which can clog the homogenizing valve.

6.3.2.4 *Sonication.* Frequencies above the limit of human hearing (ultrasonics, of frequency 20 kHz and above) applied to solutions cause 'gaseous cavitation', i.e. areas of rarefaction and compression which rapidly interchange. On the collapse of gas bubbles produced in cavities, shock waves are formed: these constitute the destructive element of this procedure. Sonication in batch or continuous processing has been employed successfully for the disruption of many types of microbial cells but others, e.g. *Staphylococcus*, are resistant. Success is dependent on the correct choice of pH, temperature and ionic strength. The selection of these parameters is empirical and will vary with both the organism and the product required.

Ultrasonic treatment is not satisfactory for large-scale use but is a versatile method for laboratory-scale work. Here too, heating is a hazard, and free radical production may also cause enzyme inactivation.

6.3.2.5 *Solid shear.* A combination of the advantages given by agitation with abrasives and by liquid shear is achieved by forcing frozen cell pastes through an orifice. Disruption is achieved in such equipment by shear forces exerted by passage through the orifice aided by the ice crystals in the frozen paste. There is no obvious hazard to the enzyme(s) in this treatment, since the process uses a temperature of $-20°C$. The principle of solid shear has apparently not been applied to the large-scale extraction of enzymes.

6.3.2.6 *Freezing and thawing.* If cell pastes are stored at $-20°C$ they will inevitably be subjected to a thawing stage before treatment. The formation and melting of ice crystals disrupts the cellular material sufficiently to allow the release of some proteins, about 50% of the proteins of the periplasmic space but only 10% of the total soluble protein. Thus the technique is not generally applicable, but if a specific enzyme can be released by this means in reasonable yield, that enzyme should be of high purity.

6.3.2.7 *Osmotic shock.* Enzymes and non-enzymic proteins may be released from the more delicate Gram-negative bacteria by suspending cells in a buffered solution of high osmotic pressure, such as 20% sucrose, then, after equilibration and concentration by centrifugation, suspending them in water. The resulting shock due to entry of water, which is hardly sufficient to be called 'bursting', allows the release of 4–7% of the total bacterial protein from the more delicate bacteria. As with freezing and thawing, enzymes released may well have been purified some 15-fold relative to the total protein content of the bacterium. Gram-positive bacteria are not affected by osmotic shock, probably because the osmotic pressure of their contents is normally very high, making them more able to resist such bursting pressures.

6.3.3 Extraction by chemical methods

The more useful physical extraction methods are relatively violent and may cause problems by comminuting cell walls, making subsequent centrifugation or filtration difficult, or by damaging the enzymes themselves. Chemical extraction allows very gentle release of enzymes, and only brief gentle agitation is required to disperse the extracting chemical during the extraction process.

6.3.3.1 *Detergents.* Anionic (e.g. sodium lauryl sulphate), cationic (e.g. cetyltrimethylammonium bromide) and nonionic (e.g. Tweens, Spans, Tritons) detergents are available and suitable for enzyme extraction. In appropriate conditions of pH and ionic strength, detergents combine with lipoproteins to form micelles, and therefore the lipoprotein constituents of biological membranes can be solubilized and enzymes released. Such techniques have been used to release cytochrome oxidase from beef heart, and cholesterol oxidase from *Nocardia* cells. However, detergents are likely to inactivate some enzymes by causing protein denaturation and precipitation and are therefore less than ideal extracting agents. Detergents may have to be removed before application of the enzyme, although in some cases traces of detergent can stabilize the enzyme.

6.3.3.2 *Lytic enzymes.* Enzymes provide very gentle and selective means of disrupting microbial cells. The best-known lytic enzyme is lysozyme, produced in commercial quantities from egg-white, which catalyses the hydrolysis of the β-1,4-glycosidic bonds in the polysaccharide portion of bacterial cell walls. Polysaccharide is more important structurally in Gram-positive than in Gram-negative bacteria and consequently these are more susceptible to disruption by lysozyme. Slight osmotic shock is necessary to rupture the cell membrane once the cell wall has been removed. The addition of EDTA (ethylene diamine tetra-acetic acid) to chelate divalent cations is necessary before lysozyme can be used to disrupt Gram-negative cells.

Similar lytic enzymes are known which may be used to release enzymes from yeast and fungal cells. Drawbacks include their cost and the likely need to remove them later in the enzyme purification process. Nevertheless, lysozome has been used commercially in the extraction of glucose isomerase from cells of *Streptomyces* species.

6.3.3.3 *Alkali.* Brief treatment (20 min) at pH 11–12.5 causes hydrolysis of cell wall material and enzyme release. This approach proved suitable for the extraction of L-asparaginase from *Erwinia* cells. Plainly, success depends on the stability of some enzymes to high pH.

For a more detailed review of methods of cell disruption, see Chisti and Moo-Young (1986).

6.4 Primary clarification of the soluble enzyme

The solubilized enzyme, or any enzyme produced extracellularly, exists in the presence of the cells (or their debris) which produced it, together with many other enzymes and proteins, nucleic acids, salts, other low molecular weight metabolites, medium components and extracting materials where used.

6.4.1 *Centrifugation*

On the laboratory scale, centrifugation (see Table 6.2) is the primary clarification method most often selected because effective equipment is usually readily available. On a larger scale, centrifugation is not so satisfactory. One reason is the lack of large-capacity, high-speed ultracentrifuges, but there are others, discussed below.

Table 6.2 Performance of various types of centrifuge.

Rate of throughput $\phi = \dfrac{d^2(\rho_s - \rho_e)g}{18\eta} \cdot \dfrac{\omega^2 rv}{Sg}$ (for symbols, see text).

Centrifuge factors			Medium (liquid) factors	
Centrifuge type	Angular velocity ω	Rotation radius r	Volume v	Thickness of liquid layer S
Tubular bowl	High	Small	Medium	Small
Multichamber and disc	Low	Large	Large	Small
Solid bowl/scroll	Low	Medium	Large	Medium
Laboratory angle/swing-out	High	Large	Small	Large

6.4.1.1 *Theory.* The throughput of a centrifuge may be described by

$$\phi = \frac{d^2(\rho_s - \rho_e)g}{18\eta} \cdot \frac{\omega^2 rv}{Sg}$$

where ϕ = the throughput for complete particle removal
d = particle diameter
ρ_s = particle density
ρ_e = fluid density

g = force of gravity
η = kinematic viscosity
ω = angular velocity
r = rotation radius
v = volume of liquid in centrifuge
S = thickness of liquid layer in centrifuge

Of the right-hand side of the expression, the first component describes the terminal velocity of the particle of diameter d under the influence of gravity; the second relates to the characteristics of the centrifuge.

Efficient operation is favoured greatly by large particle diameter d, large density difference between densities of particle and fluid, and low fluid viscosity. High angular velocity, large centrifuge radius and thin sedimentation layer also assist. In practice, particles of biological material are often small and of low density, while fermentation broths are often viscous and occasionally of fairly high density. On the laboratory scale, high angular velocities may be used, but only in batch use. Ideally, industrial centrifuges must be capable of running continuously: the radius of rotation r is restricted (mechanical stress increases with r^2 and safety limits are reached rapidly), and the need for continuous flow-through of liquids restricts the angular velocity.

6.4.1.2 *Tubular bowl centrifuge.* This is essentially a cylindrical tube driven by an overhead power unit connecting to a flexible shaft. Very high angular velocities (ω) may be employed—up to 50 000 rpm is possible in one Sharples model. The material to be clarified is pumped into the bottom of the rotating cylinder, and the thickness of the liquid layer S is small, as is the volume v, so precipitation of cell debris occurs readily if the medium viscosity is not high. Liquid flow is continuous in operation, but the centrifuge must be stopped occasionally to empty out the precipitate. Hazards to the enzyme are aeration and the consequent foaming of the clarified solution, while aerosol formation may be a hazard to the user.

6.4.1.3 *Multichamber and disc centrifuges.* These have a comparatively large radius, and consequently are capable only of low angular velocity, but are arranged so as to have multiple chambers in which the liquid layer is thin. Thus sedimentation occurs with high efficiency over large surface areas. Batch and continuous discharge machines are available, the latter having the drawback that the sedimented material must be comparatively moist to flow out of the bowl. Disc bowl centrifuges are widely used, for example to harvest baker's yeast. As the bowl is located above the gearbox, some heating is inevitable, representing a danger to enzymes.

6.4.1.4 *Solid bowl scroll centrifuge.* This comprises an Archimedean screw inside a solid bowl. As the two components rotate at slightly different speeds, sedimented solids are scrolled along the bowl: thus liquids leave at one end, solids at the other. For mechanical safety, rotation speeds are comparatively low, therefore only coarse solids may be sedimented effectively, but there are few hazards to the enzymes.

6.4.1.5 *Perforate bowl basket centrifuge.* This class of centrifuge is best illustrated by the domestic spin-drier. Fine precipitates will blind the perforate bowl, but the basket centrifuge is an excellent means of removing coarse debris from plant homogenates for example. Laboratory-scale basket centrifuges are dangerous to use and are no longer available, but large machines, some feet in radius, are widely used in industry.

6.4.2 *Flocculation and coagulation*

Centrifugation and filtration are aided by aggregation of the particles. Flocculation occurs when an agent, often in very dilute solution, bridges particles to produce a loose aggregate. The flocculating agent may be present from the beginning of the process or, more usually, be added at some suitable time. Coagulation is the direct adhesion of particles to one another, as when their charge is neutralized and coalescence is possible. These techniques may be applied to whole cells, cell debris and soluble proteins.

6.4.2.1 *Whole cells.* Many natural and synthetic polymers, such as gelatine or polyacrylamides containing various proportions of acrylic acid, may be used to flocculate microbial cells, as may inorganic salts such as alum, ferric and calcium salts. True flocculation is a very efficient process, concentrations of flocculant as low as 3 ppm being effective. Care is necessary in the choice of flocculant as it must not inhibit enzyme activity or interfere with subsequent processes such as cell immobilization. Where high molecular weight polymers are employed as flocculants they are likely to be damaged irrevocably by quite gentle shearing forces, so the conditions of addition to the cell suspension are critical. Most flocculants have a fairly clearly defined effective concentration: further addition of flocculant is countereffective.

The state of the cell surface is critically important in the selection of flocculating agents. The charge on cell surfaces is normally negative, yet strong anionic polymers may be effective because of interaction with adsorbed medium components. Washed cells may require different concentrations of flocculating agent, or even different flocculants.

Flocculation is a cheap and effective process which allows the use of cheaper centrifuges or even their elimination and is being employed increasingly in industrial fermentations.

6.4.2.2 *Cell debris and protein.* Before intracellular enzymes can be recovered from disrupted cell material the cell wall debris must be removed. Flocculation or coagulation are potentially useful techniques for removing small particles, but the materials used must not make the enzymes insoluble.

Soluble enzymes may be coagulated using tannic acid, but this treatment may inactivate certain enzymes. Tannic acid at 0.1–1.0% gives filterable precipitates, but these may be difficult to redissolve.

6.4.3 *Filtration*

The rate of passage of a liquid through a filter of unit area is dependent on the pressure difference applied, the resistance of the filter material, the viscosity of the liquid, and the resistance produced by cake already present. Thus the effectiveness of a filter will initially be high, but will fall as material accumulates and perhaps compresses. *Filter aids*, such as diatomaceous earth, retain finer particles and are valuable in enzyme isolation, but they tend to occlude liquor containing the enzyme and will damage downstream

Table 6.3 Methods of concentrating enzymes.

Basic technique	Specific agent	Effectiveness	Drawback
Precipitation			
(a) Inorganic salts	Ammonium sulphate	***	
	Sodium chloride	*	
	Sodium sulphate	**	
	Calcium acetate	*	
(b) Organic solvents	Acetone	(**)	Fire hazard
	Ethanol	(**)	
	Isopropanol	(**)	
(c) Charged polymers	Cetavlon	**	
	Protamine sulphate	(**)	Cost
	Polyethyleneimine	**	
	DEAE-Dextran	*	
(d) Uncharged polymers	Polyethylene glycol	**	
Adsorption			
(a) Anionic polysaccharides	DEAE, QAE, cellulose etc.	***	
(b) Cationic polysaccharides	CM, SP, phosphocellulose	***	
(c) Affinity chromatography		****	Cost
(d) Others	Alumina	*	
	Celite	*	
	Calcium phosphate	**	
Ultrafiltration reverse osmosis		***	
Drying			
	Rotary evaporation	*	Heat
	Spray drying	*	
	Freeze drying	***	
Freezing		**	

Key: **** Excellent
  *** Very good
  ** Good
  * Fair

equipment if allowed to pass into the filtrate. Siliceous materials constitute a health hazard and must be treated with care, particularly when dry.

The commonest forms of industrial filter are the plate and frame press, and the rotary drum filter. The former consists of filter cloths trapped between corrugated plates; fluid passes in at one side of the cloth and out, via the corrugations, to a pipe serving all the units in the battery. To remove the solids the plates must be parted manually or semi-automatically and the cloths thoroughly cleaned. Fungal mycelia are readily removed in this manner, as are bacteria after flocculation and enzymes precipitated by ammonium sulphate.

In the rotary drum filter, vacuum is applied to the inside of a hollow drum rotating in a trough containing the material to be filtered. Sediment accumulates on a filter cloth from which it may be removed by a multiplicity of methods.

6.5 Concentration

Enzymes in solution may be concentrated by a variety of means, summarized in Table 6.3.

6.5.1 Removal of nucleic acids

Nucleic acids may be removed from cell-free extracts by a variety of means. An obvious and inexpensive technique is enzyme hydrolysis using nucleases, which is applicable except, perhaps, where the enzyme(s) required are themselves nucleases. Otherwise, the nucleic acids may be precipitated using high molecular weight cations such as polyethyleneimine, streptomycin sulphate, cetyltrimethylammonium bromide or protamine sulphate. Manganese chloride is sometimes recommended, but is relatively inefficient.

6.5.2 Precipitation

6.5.2.1 *Ammonium sulphate.* Enzymes may be precipitated and fractionated by 'salting out', usually by ammonium sulphate which is cheap, very soluble, self-cooling on dissolving in water, and harmless to most enzymes. The logarithm of the decrease of protein solubility in concentrated salt solutions is a linear function of increasing ionic strength, the general equation being

$$\log s = B' - K'_s \frac{\Gamma}{2}$$

where s = solubility of protein in g l^{-1} of solution
 B' = intercept constant
 K'_s = salting-out constant
 Γ = ionic strength (moles l^{-1}).

B' varies with pH, temperature, and the properties of the enzyme in solution. K'_s is independent of pH and temperature but will vary with the salt used and the enzyme.

A given protein at concentration s will begin to precipitate at the ionic strength given by

$$\frac{\Gamma}{2} = \frac{B' - \log s}{K'_s}.$$

Clearly, precipitation of an enzyme will be aided if its intrinsic solubility is at a minimum, i.e. at its isoelectric point, the pH where there is no net charge on the molecule. In general the majority of proteins are precipitated at an ammonium sulphate concentration that lies between about 15% and 40% (w/v). In many publications, ammonium sulphate concentrations are described as % of saturation rather than % w/v dissolved, a confusing practice, best avoided.

6.5.2.2 *Organic solvents.* Organic solvents reduce the dielectric constants of aqueous media, and thus may lower the solubility of proteins by allowing the protein molecules to interact more readily with each other than with the water. Alternatively, solvent molecules may exchange with protein-bound water, or cause precipitation simply by dehydrating the protein. Probably all three effects operate in protein precipitation.

Proteins can be precipitated very satisfactorily using organic solvents provided that the temperature is below 4°C; above this, denaturation can occur. Suitable solvents must be miscible with water and should have low flammability: isopropanol for example has been used to precipitate amyloglucosidase. Methanol, ethanol and acetone have also been used, but their industrial use is limited because they are inflammable and relatively costly.

6.5.2.3 *High molecular weight polymers.* Polyethylene glycol may be used very satisfactorily to precipitate proteins. Unlike solvents, it has a protein-stabilizing effect and thus may be used at ambient temperatures. It is effective at relatively low concentrations, most proteins being precipitated at polyethylene glycol concentrations of 6–12%, and thus 50% w/w aqueous solutions of the precipitant (molecular weight 6000) may be used very conveniently. Protein solubility with respect to polyethylene glycol is influenced by pH, ionic strength, and temperature. In contrast to salting-out, precipitation by polyethylene glycol depends on protein concentration.

An alternative means of concentrating proteins is to use polyethylene glycol to withdraw water from the protein solution through a dialysis membrane.

6.5.3 *Ultrafiltration and reverse osmosis*
Reverse osmosis/ultrafiltration is a means of concentrating enzymes which is

rapidly gaining popularity. In ultrafiltration, molecules are forced hydraulically through a membrane of very small pore size. 'Reverse osmosis' is simply ultrafiltration using a membrane with pores small enough to allow the passage of solvent molecules only, as in the desalination of sea water. Reverse osmosis may thus be used to concentrate enzyme solutions. Slightly less fine-pored membranes can be used for low molecular weight solutes, as in the removal of ammonium sulphate after enzyme precipitation. Ideally, ultrafiltration can fractionate proteins on a molecular weight basis, but this is rarely practical on a large scale because membrane polarization difficulties cause blocking (see below).

Ultrafiltration membranes are available with molecular weight cut-offs between 500 and 300 000. There are two types of ultrafiltration membrane, microporous and diffusive. The former is a rigid membrane with small pores running through it of average diameter 500–5000 Å. Very small molecules will pass through the membrane and large ones will be retained at the filter surface. Intermediate-sized molecules will be retained within the structure of the membrane and will eventually block the pores. For this reason microporous filters have been superseded by diffusive membranes.

Diffusive ultrafilters consist of homogeneous, hydrogel membranes through which solvents and solutes are transported, under the influence of a concentration or activity gradient, by molecular diffusion. Thus the transport of a molecule across the membrane requires kinetic energy and occurs more readily at high temperatures. A readily permeable membrane would be readily hydrated and composed of a polymer with a strong specific affinity for its solvent. Conversely, a relatively unhydrated, rigid membrane would have low permeability, particularly when there was reduced affinity between the membrane polymer and the solvent.

In practice, anisotropic diffusive membranes are used. These consist of very thin (0.1–5 μm) active skins supported by a substructure (20 μm–1 mm). The support material porosity is chosen so that it will not retain any molecule that passes through the membrane. Such a combination of active membrane and support material should not be blocked by solutes during use. Solute flux is governed by the laws of diffusion within the membrane, i.e. is independent of pressure but determined by solute concentrations in the concentrate and permeate. Hence if pressure is increased, increasing solvent flux, the retention efficiency of the membrane increases because the solute flux is barely altered. Solute diffusion coefficients differ with varying shapes and sizes of molecules (thus globular proteins and linear polysaccharides of similar molecular weights will diffuse at very different rates) but in general they vary exponentially with the square of the molecular diameter.

'Membrane polarization' is a phenomenon arising as a build-up of rejected solute at the membrane surface. This impedes solvent flux so that eventually it may no longer respond to an increase in hydraulic pressure. Thus the design of ultrafiltration apparatus is aimed at minimizing the build-up of such polarization layers, in small scale apparatus by stirring, and in larger equipment by maintaining high fluid flow rates over the membrane surface. Reduced membrane efficiency may result from the adsorption of proteins. This may be overcome by brief exposure to alkali or by treatment with proteases.

Ultrafiltration is a particularly useful method for enzyme work because:

(i) It is gentle and non-destructive (though shear damage of large molecules is possible)
(ii) No chemical reagents are used
(iii) No phase change is required
(iv) It can be operated at low hydrostatic pressures
(v) Low temperature operation is possible (though cooling is necessary in equipment where high flow rates are maintained over membranes)
(vi) Constant pH and ionic strength may be maintained if required
(vii) Simultaneous concentration and purification can be achieved
(viii) The process is economical in use.

6.5.4 *Freeze-drying*

Freeze-drying (lyophilization) relies on the ability of ice to sublime. At reduced pressures with a suitably cold sink, the process occurs rapidly and solutes are dried very effectively, remaining as fluffy, readily soluble powders. To be effective in practice for protein-concentration and drying, the protein solution must be virtually free of low-molecular weight solutes, otherwise eutectic mixtures will form. These will foam, causing physical loss of protein and denaturation.

The process has been used widely in preparing many enzymes commercially, but cannot be used in all cases because some enzymes, in particular those composed of sub-units, may be completely inactivated by freeze-drying.

6.5.5 *Evaporation*

Simple evaporation is a fairly useful technique for those enzymes which are not harmed by increased concentrations of low molecular weight solutes, and are also stable to the somewhat elevated temperatures which are necessary for the process to occur at a practically useful rate. As operation under vacuum is necessary, foaming of protein solutions is always a potential hazard: such surface effects are a cause of enzyme denaturation.

6.5.6 *Freezing*

In theory, it is possible to concentrate any water-soluble substance (which does not form a solid solution with water) by freezing the water and separating the ice crystals from the remaining solution. Plainly, simply freezing the bulk solution would ultimately give a useless block of ice containing the enzyme. The conditions used must allow the formation of ice crystals which do *not* contain the enzyme in inclusions or occlusions: this is best done by stirring the solution and allowing ice crystals to grow reasonably slowly.

Once this is done, the ice crystals may be removed from the liquid phase by centrifugation or filtration, but this does not result in perfect separation, since part of the liquid enzyme concentrate adheres to the surface of the crystals. The adhering material may be washed off with the least wastage using melted pure ice crystals.

The technique has proved very successful in concentrating culture filtrates of various organisms with very little loss of enzyme activity. In spite of this it has not been used extensively, but given further development it could be a powerful tool in enzyme purification.

6.6 Enzyme purification—chromatography

The techniques described above are primarily concerned with enzyme recovery, but may be used with limited success for enzyme purification by protein fractionation. Purification is best achieved, however, by chromatography and related techniques.

The basis of purification by chromatography is the retardation of solute molecules during the passage of a solution through a column containing particles of solid material. Since retardation, rather than permanent binding, is required, the material to be purified must be capable of 'binding' to the solid phase so as to be retarded, and of re-entering the solution later, so as to be eluted. In other words, it must be able to partition itself between the solid phase and the liquid phase. Table 6.4 summarizes the basis of enzyme separation and purification by the major chromatographic techniques.

6.6.1 *Gel chromatography*

This technique, well known on the laboratory scale as a means of determining the molecular weight of proteins, serves also for their purification. The materials most frequently encountered are the various grades of Sephadex (cross-linked dextrans), Sepharose (agaroses) and Bio-Rad gels (polyacrylamides), among many others. Many different expressions have been used for the same principle: gel filtration, molecular sieve chromatography, exclusion chromatography, gel permeation chromatography, and gel chromatography are one and the same. Together, these names indicate the principle of the process.

Sephadex gels consists of dextran molecules cross-linked with epichlor-

Table 6.4 Types of chromatography.

Type	Basis of separation	Basis of fractionation	Means of solute elution
Gel chromatography	Partition between liquid in pores of stationary phase and the bulk liquid	Molecular size	Continued flow of initial solvent
Ion exchange	Electrostatic binding between solute and stationary phase	Net charge on solute and ion-exchanger	Change of ionic strength or pH or both
Affinity chromatography	Binding between biologically active molecule and an immobilized specific ligand, e.g. substrates, co-factors (NAD, AMP), dyes lectins, antibodies	Binding or not	Ligand with higher specificity for enzyme than column Change of ionic strength pH change Chaotropic agents etc. etc.

hydrin. The amount of epichlorhydrin used determines the extent of cross-linking which in turn determines the degree of water regain which is possible. The greater the water regain the greater the porosity, i.e. the larger the molecular species which can be fractionated. Since the porosity may be determined both by dextran concentration and by the extent of cross-linking, a wide range of porosities is possible.

Bio-Gels consist of beads of polyacrylamide which again can be manufactured with a range of porosities determined by the initial concentrations of monomers and by the amount of cross-linking.

Agarose gels have larger pore sizes than dextran or polyacrylamide gels, and therefore are capable of fractionating larger proteins. They are comparatively unstable, though, so composites of agarose with polyacrylamide (Ultrogels) and agarose cross-linked with 2,3-dibromopropanol (Sepharose CL) have been developed to retain both the desirable properties of agarose and a useful working life.

If a mixture of molecules of different sizes is placed at the top of a column of gel, the largest molecules will be too large to enter the pores and will therefore be excluded and pass through the column unretarded. The lower molecular weight materials, however, will be retarded by passing through the stationary phase inside the pores of the gels. Thus gel chromatography is the diffusional partitioning of solute molecules between the solvent phase and the solvent confined in the spaces within the porous beads of the stationary phase.

The total volume of a gel column or bed is

$$V_t = V_0 + V_i + V_g$$

where V_g = volume of gel matrix
V_i = volume inside bead
V_0 = volume outside bead.

A molecule capable of entering the pores of a gel will have a partition coefficient representing the extent to which it can enter the stationary phase. The partition coefficient, K_{av}, is given by the equation

$$K_{av} = \frac{(V_e - V_0)}{(V_t - V_0)}$$

where V_e = volume of solvent required to elute the solute from gel column or bed (experimentally determined)
V_t = total volume of gel column or bed
V_0 = void volume (the experimentally determined volume outside the bed).

It has been shown empirically that K_{av} is inversely proportional to the logarithm of the molecular weight of solute, at least for globular proteins. Each type of gel will be capable of excluding molecules larger than a particular size, and therefore of fractionating molecules within a particular range and of desalting solutions of proteins.

Most gels contain a low content of ionic groups which may lead to ion-exchange effects. These can be overcome by equilibrating the gels in buffers with a concentration of over 0.02 M.

6.6.2 Ion-exchange chromatography

Very many types of ion exchange material are available, some of which are unsuitable for enzyme purification. They fall into two general classes, ion exchange resins and ion exchangers based on hydrophilic organic polymers such as cellulose, dextran, agarose and polyacrylamide.

6.6.2.1 *Ion exchange resins*. These are water-insoluble polymers containing anionic or cationic groups as $-SO_3^-$, $-COO^-$, quaternary ammonium or polyamine residues. They often have high capacities compared with ion exchange celluloses and are designed robustly for repeated use in columns, in particular for deionizing water. Their high capacity means that they are potentially very useful for protein concentration and purification. However, extremes of pH or ionic strength are required to elute proteins from these resins, and may cause denaturation. Ion-exchange resins are worth considering for the purification of enzymes that are known to be robust.

6.6.2.2 *Ion exchangers based on hydrophilic polymers.* Cellulose and the various materials used for gel chromatography can be derivatized chemically to introduce various charged groups, anionic or cationic. The level of derivatization is variable, but for cellulose-based materials a maximum of 2.0 meq g^{-1} employed because:

(i) This low level of derivatization allows binding and elution of polyelectrolytes in mild conditions and has been found to be satisfactory in practice
(ii) Higher levels of derivatization may solubilize the cellulose.

The gel chromatography materials are cross-linked so are not vulnerable to solubilization, and high degrees of substitution can be achieved without breakdown of the gel. In practice the capacities used are between 2.3 and 4.5 meq g^{-1}.

The most commonly-used ion exchanger of this type are diethylaminoethyl- (DEAE-) and carboxymethyl- (CM-) cellulose, Sephadex etc. Sulphopropyl (SP) and quaternary aminoethyl (QAE) diethyl-(2-hydroxypropyl) aminoethyl Sephadex and cellulose are also available. DEAE and CM are weak anion and cation exchangers, respectively: QAE and SP are strong anion and cation exchangers respectively.

6.6.2.3 *Protein purification on ion exchangers.* The basis of protein purification using ion exchangers is the reversible adsorption of proteins by mainly electrostatic forces. The magnitude of these forces may be changed by altering the pH of the environment, which will change the net charge on the proteins and the extent of dissociation of weakly acidic or basic ion exchangers, or by changing the ionic strength of the environment. Increasing the ionic strength increases the competition for binding sites in the ion exchangers and thus changes the partition coefficients of proteins, favouring their removal from the ion exchangers. Thus proteins may be adsorbed tightly to ion exchangers and desorbed by changing pH or ionic strength or both. Desorption may be done batchwise or continuously, in which case chromatography is occurring. Proteins may be separated very effectively by column chromatography using gradients of pH or ionic strength. Equipment is available which allows the use of complex gradients to give maximum resolution of mixtures of proteins.

In more detail, proteins are adsorbed on to cationic exchangers (R—COO^{-}) in conditions of low ionic strength and at a pH below the protein's isoelectric point, i.e. when the net charge on the protein is positive (P—NH$_3^+$). For adsorption to anion exchangers the situation is reversed (but ionic strength is still kept low). The weak ion exchangers DEAE- and CM-cellulose, etc., are most often used. Adsorption to those is usually a rapid process, complete in 10–20 minutes, but some enzymes are adsorbed much more slowly. Use of the strong ion exchangers QAE- or SP-Sephadex may overcome this problem.

In batchwise use, the ion exchanger is equilibrated with a buffer solution selected to allow adsorption of the enzyme(s) to be purified, then stirred with the sample until adsorption is complete. It is then collected in a suitable filter, unbound materials are washed away, and the enzymes are eluted using appropriate changes of pH or ionic strength. Where the isoelectric point of an enzyme is extreme, batchwise use of ion-exchangers can be a very potent method of purification, either by binding all the unwanted proteins but not the required enzymes, or vice versa. Column chromatography is a more generally useful purification process, but once the characteristics of the enzyme to be purified are known, and especially if a high degree of purification is not needed, the batchwise use of ion exchangers is a cheap and rapid method which can readily be scaled up.

6.6.3 *Affinity purification*
This is an extremely powerful technique which can be highly specific for individual enzymes (e.g. Lowe, 1981). In the most specific (and usually expensive) form of affinity purification, an analogue (which might be a competitive inhibitor) of the substrate or cofactor of an enzyme is attached to a support material and used as a ligand. On contacting this with a mixture containing the enzyme, an enzyme–ligand complex forms. This is highly specific, and no other proteins in the mixture will be bound. The enzyme may be released from the complex by treatment with a solution of the normal substrate of the enzyme or by changing the pH sufficiently to change the conformation of the enzyme and break the complex. Specific antibodies are equally effective as ligands for enzyme purification.

For large-scale use, the application of substrate or cofactor analogues or of antibodies as ligands for enzyme purification can be prohibitively expensive. A wide range of alternative ligands have been developed which are less specific but still utilize the principle of affinity with a part of the enzyme molecule. Many water-soluble dyes act as ligands with quite high specificities for enzymes (Hey and Dean, 1981). Lectins may be used for the purification of glycoprotein enzymes. Other enzymes may be purified using their interaction with hydrophobic groups or metal chelates (Janson, 1984). It is reasonable to assume that any enzyme could be purified using an affinity method.

Ligands for affinity purification must be covalently bound to a support material in such a way that there is minimal steric hindrance of access of the enzyme to the ligand. It is essential that the ligand–enzyme complex can be broken when required: some competitive inhibitors may bind so tightly that the enzyme would be inactivated by the conditions needed to release it. For maximal effectiveness the support material must not be capable of binding other proteins non-specifically. Gels such as agarose and dextran are particularly suitable as support materials. They must be activated, for example by cyanogen bromide, and the extent of activation should be great to give high adsorptive capacity. To minimize steric hindrance, ligands are attached to supports via flexible 'spacer arms', usually hydrocarbon chains 6–9 carbon atoms long (Lowe, 1981).

At their best, affinity purification methods are capable of purifying enzymes and other valuable proteins several hundredfold in a single step. For example, human interferon can be isolated using a monoclonal antibody in a single step that achieves a 5000-fold purification.

6.6.4 *Chromatography columns*

The efficiency of a chromatographic system is judged by its ability to resolve mixtures of similar materials (or to allow the elution of a single component in the minimum volume of eluting fluid). Resolution depends on the differential rate of migration of zones of proteins and on the rate of spreading of the zones, which opposes the effect of the differential migration rates. The magnitude of the differential rate of migration is determined by the nature of the proteins and of the stationary and liquid phases. Table 6.5 summarizes the causes of zone spreading and indicates how they may be reduced. Clearly zone spreading cannot be eliminated but may be minimized by the choice of a compromise of column parameters. The most satisfactory columns available for large-scale gel chromatography consist of multiple short wide units connected in a 'stack'.

For ion-exchange and affinity chromatography column design is less critical, although it should not be forgotten that ion exchangers undergo volume changes, sometimes quite dramatic, in use.

Table 6.5 Zone spreading in gel chromatography.

Cause	Counteraction
Normal diffusion	Speed-up separation Short (wide) columns
Trailing caused by insufficient equilibration time	Slow down separation Short columns
Eddy diffusion in areas where differences in flow rate are possible	Assure even bed packing, uniform particle size
Channelling	Long narrow columns Avoid bed compression
Uneven application of material to solid phase bed. Uneven reception from bottom of bed.	Good design of end pieces

6.6.5 HPLC of enzymes

The recent very rapid development of methods for the production of chromatographic media in the form of small, uniformly-sized beads has allowed the application of high performance liquid chromatography (HPLC) to the separation and purification of proteins. Pharmacia's trademark, FPLC (fast protein liquid chromatography), tends to be used more widely than that company intended for the HPLC of proteins. Ion exchangers of various types, gel filtration media, hydrophobic media and affinity media can all be obtained in forms suitable for use in HPLC equipment. The principles remain the same as in conventional liquid chromatography, but HPLC is a more powerful technique because of the speed and convenience with which fast, high-resolution separations may be achieved. Using HPLC, complete separations of mixtures of proteins can be achieved in minutes rather than the hours needed for conventional chromatography. This is greatly advantageous in the purification of very unstable enzymes. The resolution of mixtures is usually far superior to that obtained using conventional liquid chromatography, and consequently high degrees of purification may be achieved in a single HPLC step.

At present only small columns are readily available for the HPLC of proteins, so only small quantities of enzymes may be processed in a single run, but the speed of the technique is such that useful quantities of enzymes may be accumulated from multiple runs. Preparative-scale protein HPLC equipment, to produce gram quantities of proteins, will surely be readily available in the near future.

References

Ahern, T. J. and Klibanov, A. M. (1985) *Science* **228**, 1280–1284.
Aunstrup, K. (1974) In *Applied Biochemistry and Bioengineering*, vol. 2, *Enzyme Technology*, eds. L. B. Wingard, Jr., E. Katchalski-Katzir and L. Goldstein, Academic Press, New York, Chapter 2.
Chisti, Y. and Moo-Young, M. (1986) *Enzyme Microb. Technol.* **8**, 194–204.
Darbyshire, J. (1981) In *Topics in Enzyme and Fermentation Biotechnology*, vol. 5, ed. A. Wiseman, Ellis Horwood, Chichester, Chapter 3.
Hey, Y. and Dean, P. D. G. (1981) *Chem. and Ind.* 72–732.
Janson, J. C. (1984) *Trends in Biotechnol.* **2**, 31–38.
Lilly, M. D. (1979) In *Applied Biochemistry and Bioengineering*, vol. 2, *Enzyme Technology*, eds. L. B. Wingard, Jr., E. Katchalski-Katzir and L. Goldstein, Academic Press, New York, Chapter 1.
Lowe, C. R. (1981) In *Topics in Enzyme and Fermentation Biotechnology*, vol. 5, ed. A. Wiseman, Ellis Horwood, Chichester, Chapter 2.
Tombs, M. P. (1985) *J. appl. Biochem.* **7**, 3–24.
Wang, D. I. C., Cooney, C. L., Demain, A. L., Dunnill, P., Humphrey, A. E. and Lilly, M. D. (1979) *Fermentation and Enzyme Technology*, John Wiley, New York, Chapter 12.

7 The application of immobilized enzymes and cells and biochemical reactors in biotechnology—principles of enzyme engineering
PETER S. J. CHEETHAM

7.1 Introduction

This chapter discusses the advantages and disadvantages of using immobilized enzymes or cells (immobilized biocatalysts) as industrial catalysts. The various types of biocatalyst are described, and also the methodologies which facilitate their use on a large scale, such as immobilization and the use of enzyme reactors. Various methods of immobilization are available, and a number of factors may modify the intrinsic activities of the immobilized biocatalysts, such as diffusional restrictions on the rate of supply of substate. Immobilization, the use of concentrated substrate solutions in reactors, and the need to achieve high yields of product all affect the activity, stability and selectivity of the enzyme and cell biocatalysts. Therefore the advantages and disadvantages of the various types of biological reactor need to be considered, especially the batch, continuous stirred and plug-flow types, since factors such as the reactor configuration and non-ideal flow of reactant solutions influence reactor kinetics and performance. Other factors affect the stability of biocatalysts and the possibility of regenerating the activity of biological reactors. Further important areas covered in this chapter are methods of maintaining a constant productivity from biocatalyst reactors, the mechanical problems associated with immobilized biocatalysts, and finally the methods available for the recovery and purification of the products from biological reactors. Some indication of the commercial applications of biocatalysts in various industries such as the food, pharmaceutical, chemical and waste-treatment industries are given.

7.2 The application of biological catalysts

Soluble enzymes have acquired many important commercial uses, such as the microbial proteases used in washing powders. In many cases the use of biological catalysts, in the form of enzymes of cells, is greatly facilitated by their immobilization and subsequent use in enzyme reactors. It is impossible to separate the design of the reactor from that of the catalyst, since the properties of the immobilized enzyme or cell will greatly affect the reactor design. The best example of an immobilized enzyme used in industry is glucose isomerase, which, at present, produces several million tonnes of high fructose syrup per year, mainly in the USA.

7.3 Types of enzymic catalyst and commercial applications

Several thousand enzymes possessing different substrate specificites are now known, and a much larger number certainly still await discovery. Enzyme technology can be viewed as a struggle to obtain the advantages of enzymic catalysis by overcoming or circumventing the inherent disadvantageous features of enzymes. Several main types of biological catalyst can be identified (Figure 7.1), including cells, enzymes and organelles, used in both free and immobilized forms. Fermenting cells, and to a lesser extent soluble enzymes, have long been of interest, but the use of immobilized cells in industry is now common. The potential of non-growing whole cells to carry out biotransformations is now becoming increasingly recognized. Enzymes from microbial, rather than plant or animal, sources are usually used because of their easier availability, greater stability and their variety and ease of genetic

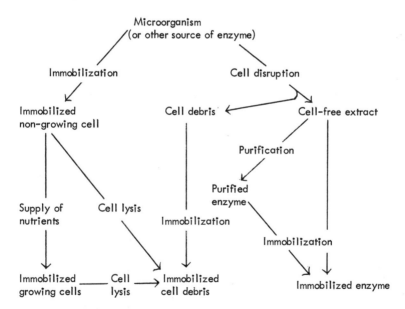

Figure 7.1 The interrelationship between the various forms of enzyme catalyst.

manipulation. Extracellular rather than intracellular or membrane-bound enzymes are often preferred because they are easier to obtain from the parent cell. Cell-associated activities are often best used in the form of immobilized cells, and there is increasing interest in the use of immobilized plant and animal cells, for instance, to produce food flavours and monoclonal antibodies respectively.

Enzymes are complex, highly active, renewable, stereospecific catalysts which act on a wide range of substrates under mild conditions in aqueous solvents. They require only low energy inputs and have only low heats of reaction, but are generally more labile than chemical catalysts, a property which can be used to advantage, as reaction can then be terminated by heat denaturation. Biocatalysts are most useful when chemical catalysts are not effective, for example when stereospecific products are required. However, a typical problem of enzymic reactions is that only partial conversion of a dilute substrate solution is possible, necessitating the separation of the desired product molecules from unreacted substrate and, particularly, very large quantities of solvent (usually water).

In the selection of a catalyst, enzymes also have to compete with chemical catalysts and with fermentation processes in which cell growth and product formation occur simultaneously. Usually, enzymes require less expensive reactors and less complex media than fermentations, aeration and sterile conditions are not required, purification of products is simpler, and less waste is produced. Continuous operation is easier, and higher yields of product are usually formed. However, because of requirements for regenerated cofactors (e.g. $NADH-NAD^+$) and energy sources such as ATP, synthetic and multistep reactions are much more difficult to perform using isolated enzymes, but can often be achieved by using immobilized whole cells. Bioconversion reactions occupy an in-between position between soluble enzyme reactions and fermentations. Bioconversions are reactions carried out by one enzyme or a small number of enzymes still associated with the cell.

New biocatalysts are usually obtained by microbiological screening techniques, often of microorganisms from environments that exert a selective pressure on the microorganism. However, a new promising development termed protein engineering (site-directed mutagenesis) promises to make any enzyme activity available by modification of the gene coding for the enzyme, and thus altering the active-site amino acids of the enzyme or features responsible for characteristics of the enzyme, such as the thermostability and resistance to denaturants.

Currently, the industrial applications of soluble enzymes include the use of alkaline proteases in washing powders and α-amylase, glucoamylase and pullulanase in the production of glucose and fructose syrups from starch (Table 7.1). Likewise, maltose syrups are produced from starch using malt, (α-

Table 7.1 Characteristics of the three main enzymes used in high fructose corn syrup manufacture (Gotfredsen et al., 1985)

	α-Amylase	Glucoamylase	Glucose isomerase
Turnover number under conditions of application (mol product/mol enzyme/minute)	3×10^4	10^4	2×10^2
Relative price of enzyme	67	8	100
Amount of enzyme used ppm of product	12	150	10–20
Relative cost of enzyme	80	133	100
Tonne product produced per kg enzyme	80	7	2.5–5

amylase) or fungal β-amylase. Microbial and animal chymosin is used for milk coagulation, lipases and proteases for accelerated ripening and flavour development in cheeses, and proteases for protein processing, recovery from scrap meat and fish, and also meat tenderization. Proteases are also used for dehairing and rehydrating hides and skins during leather manufacture. Lipoxygenase is used to bleach flour, and dextanase to hydrolyse the polymer present in some sugar cane extracts. α-Amylases are used to reduce the viscosity of the starch used to size paper, to remove the starch size from textiles, to degrade the clay-starch suspensions discharged from paper mills, and also to supplement the cereal malt traditionally used to lower the viscosity of cereal mashes in distilleries.

Combinations of proteases and amylases are used to partially depolymerize the starch and protein in flour prior to baking. The amount of free sugar formed influences the rate of gas production by the yeast, and the depolymerization of gluten by protease facilitates good dispersion of the CO_2 through the flour. Pectinases and cellulases are used to enhance the recovery of juice and flavour from fruit and to reduce the viscosity and haze of juice prior to pasteurization and concentration. Recently, specialized mixtures of enzymes have been developed, specifically designed to hydrolyse plant cell wall materials. Similarly proteases, α-amylase and β-glucanase can be used to degrade polymeric material in barley extracts prior to hopping and fermentation to produce beer, and also to degrade any residual starch, β-glucan or protein prior to pasteurization and bottling of the beer. Galactomananase is used to hydrolyse the gums present in instant coffee extracts that cause problems by increasing the viscosity of the coffee extract during concentration. Glucose oxidase is widely used as an antioxidant in foods and beverages. Invertase is used to produce liquid syrup centres in confectionery and to produce invert sugar syrups. Some of the best-known applications of biocatalysts are the use of immobilized glucose isomerase to produce high fructose syrups, immobilized penicillin acylase to produce 6-aminopenicillanic acid for semi-synthetic antibiotic manufacture, and the use of microbial enzymes such as 11 α- and β-hydroxylases as bioconversion agents for the synthesis of steroid hormones and drugs.

Other applications are constantly being proposed and developed. For instance, the manufacture of the high-intensity sweetener Aspartame from its constituent amino acids can be carried out by using the metalloproteinase thermolysin in reverse. Considerable progress has also been made in developing processes for producing L-phenylalanine for Aspartame synthesis by bioconversion. Other examples include the use of cells with hydantoinase activity for the production of *para*-hydroxy phenyglycine for use in semi-synthetic antibiotic manufacture, and the use of sulphydryl oxidase to remove 'off-flavours' from UHT-treated milk, the manufacture of acrylamide from acrylonitrile by immobilized cells possessing nitrilase activity, and the inter-esterification of fats to produce cocoa-butter substitute by immobilized lipase.

7.3.1 *Immobilized biocatalysts*

Immobilization can be defined as the process whereby the movement of enzymes, cells, organelles, etc. in space is completely or severely restricted, usually resulting in a water-insoluble form of the enzyme, although water-soluble immobilized enzymes can be formed by linking the enzyme to soluble polymers such as dextrans. Immobilized enzymes are also sometimes referred to as bound, insolubilized, supported or matrix-linked enzymes.

Immobilized enzymes occur naturally, for instance in the form of enzymes adsorbed to soil particles. Immobilized enzymes or cells are used in industry because re-use or continuous use of the biocatalyst is made possible so decreasing catalyst costs (Figure 7.2) and because the enzyme or cell is prevented from contaminating the product. Often the stability of the enzyme is increased by immobilization, so that a smaller size of reactor can be used to achieve the same productivity than using the corresponding free biocatalysts. Also, the advantages of the various types of enzyme reactor available can be more readily exploited using immobilized biocatalysts, and immobilization enables the individual enzymes or cells to be evenly distributed throughout the reactor, so ensuring an even supply of substrate to each enzyme or cell. However, some enzyme activity may be permanently lost during immobilization, due to irreversible denaturation by heat, pH changes, free radicals, etc., generated during the immobilization procedure, especially if this is achieved chemically. Immobilization is also an extra operation and adds

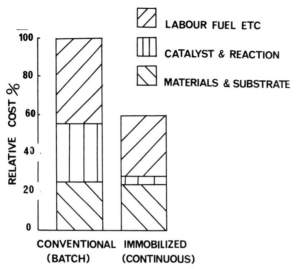

Figure 7.2 A comparison of the relative costs of the industrial production of L-aspartic acid using intact *E. coli* cells in a batch process, or immobilized cells in continuous columns. Redrawn from Chibata and Tosa (1977).

cost to a process, and the immobilized biocatalyst particles always occupy a larger volume in the reactor than do the equivalent amounts of free cells or enzyme, due to the presence of the solid support material.

Immobilized enzymes and cells are used to carry out chemical conversions, and are also used in enzyme electrodes and other analytical devices. They are proving increasingly useful as reagents in organic chemistry and in therapeutic applications, for instance as extracorporeal shunts for treatment of patients' blood. Techniques of enzyme immobilization have also been used by biochemists wishing to study the kinetics and structure of proteins, and as models of membrane-bound enzymes.

The recovery of immobilized biocatalysts, or their retention in the reactor, is achieved by the use of filters, as in packed-bed reactors; by centrifugation as in batch reactors; or by sedimentation as in the case of fluidized reactors. The most desirable situation is where the biocatalyst particles are an integral part of the reactor, as in packed columns, so that recycle loops and ancillary equipment such as centrifuges or settling tanks are not needed. Figure 7.3 represents a typical industrial immobilized enzyme/cell process. Note that the stage on which the immobilized biocatalysts is used is only a small part of the overall process, which tends to be dominated by unit operations concerned with the purification and recovery of the product.

7.3.1.1 *Methods of immobilization.* Six main approaches have been used for the immobilization of biocatalysts. These are: entrapment in polymer

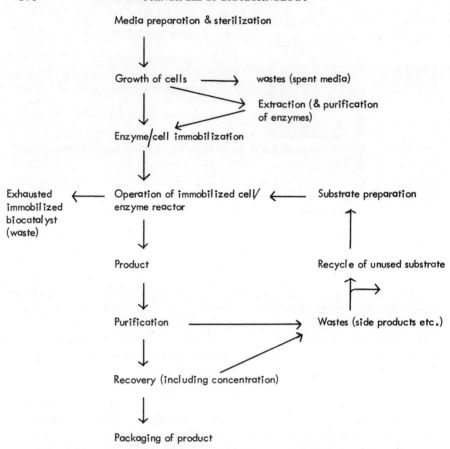

Figure 7.3 A generalized representation of a typical sequence of operations involved in industrial processes which use enzymic catalysis.

matrixes, adsorption of the charged biocatalyst on to oppositely charged support materials, covalent attachment to chemically activated supports, encapsulation inside semi-permeable membranes, aggregation of the biocatalyst particles into flocs, or biospecific attachment to supports by means of lectins, etc. Combinations of these techniques can also be used, such as adsorption of the biocatalyst to a charged support followed by cross-linking into place. There are also some methods of restricted application, such as electrodeposition on to charged supports, or immobilization of glycoproteins on to periodate-oxidized cellulose-based supports. In most cases the biocatalyst is present inside the support particle and so is present in a relatively anaerobic environment.

Covalent bonding is the most widely used method for immobilizing enzymes. The enzymes are very firmly bound, but are chemically modified and so may be denatured during immobilization. The most common technique is to activate a cellulose-based support with cyanogen bromide, which is then mixed with the enzyme. The cyanogen bromide reacts with the hydroxyl groups of the polysaccharide, and also links with amino groups on the surface of the protein. Cyanuric chloride (trichlorotriazine) can be used in a very similar fashion. However, polysaccharides are not an ideal support, as their mechanical weakness hinders large-scale use. Other supports that have been used include agar, agarose, and Sephadex; acrylate, urethane, epoxy and methyacrylate polymers; and nylon. Porous ceramic supports have proved to be especially useful support materials.

Some types of microbial cells aggregate naturally, such as the *Mortieralla* cells used in the form of pellets with α-galactosidase activity to hydrolyse oligosaccharides such as raffinose during the refining of beet sugar. Other cells and enzymes can be aggregated by cross-linking, usually by mixing with a multi-functional reagent such as glutaraldehyde, which reacts with the lysine ε-amino groups in proteins. Sometimes the enzyme is polymerized with an inert protein such as albumin. Although many other cross-linking agents, such as dimethyl adipimate, have been used, only glutaraldehyde has found widespread industrial applications, chiefly for the cross-linking of cells or cell debris containing glucose isomerase activity. Glutaraldehyde is preferred because it reacts under comparatively mild conditions and is also the only cross-linking agent permitted as a food-processing aid.

Immobilization of enzymes by adsorption is probably the mildest method available, being mediated by ionic, hydrophobic or hydrogen bonds. Manufacture of vinegar by naturally immobilized *A. aceti* cells on birch wood twigs is an established method. Adsorption is also easy to perform simply by stirring the biocatalyst with an ion-exchange resin. However, desorption often occurs after changes in pH, ionic strength or substrate concentration have occurred. Cellulose-based ion-exchange resins have been used extensively, and amino-acylase adsorbed to DEAE Sephadex is used in a Japanese industrial process for the separation of L-amino acids from the *N*-acetylated racemic mixtures produced by chemical synthesis (Figure 7.4). Adsorption effects can also be exploited by carrying out a reaction using a soluble enzyme and then recovering it by adsorption to an ion-exchange resin.

In entrapment, the enzymes or cells are not directly attached to the support surface, but simply trapped inside the polymer matrix. Thus losses of enzyme activity upon immobilization are minimized. Entrapment is carried out by mixing the biocatalyst into a monomer solution, followed by polymerization initiated by a change in temperature or by a chemical reaction. The polymer is

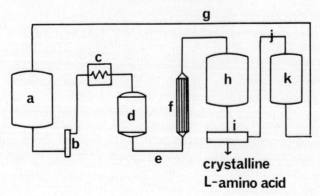

Figure 7.4 A schematic description of a continuous process for the formation and separation of L-amino acids from a racemic mixture (*N*-acetylated) using immobilized amino acylase. Redrawn from Chibata *et al.* (1976*b*). *a*, acetyl DL-amino acid reservoir; *b*, filter; *c*, heat exchanger; *d*, enzyme column; *e*, L-amino acid stream; *f*, heat exchanger; *g*, DL-acetyl-amino acid stream; *h*, crystallizer; *i*, separator; *j*, acetyl D-amino acid stream; and *k* racemization tank.

formed either in particulate form, or as a block which can be disrupted to form discrete particles. The most common methods of entrapment use polyacrylamide, collagen, cellulose acetate, calcium alginate or carrageenan as the matrices. Entrapment is chiefly used for the immobilization of cells, rather than enzymes, because their larger size means that they are more easily retained in the support. Industrial processes using *Escherichia coli* and *Brevibacterium ammoniagenes* cells, originally immobilized in polyacrylamide gel, have been developed for the production of aspartic and malic acids respectively. Subsequently, carrageenan has been used as the support material, as it has superior properties. Lactase entrapped in cellulose triacetate fibres has been used to hydrolyse the lactose in milk in Italy, glucose isomerase entrapped in collagen fibres has been used in the USA, and *Erwinia rhapontici* cells entrapped in calcium alginate gel pellets have been used to form the sucrose analogue isomaltulose.

Encapsulation envelops the cells or enzymes inside semi-permeable membranes which allow the passage of substrate and product molecules, but which do not allow the enzyme to pass through. The membranes can be formed from nylon, silastic resin or cellulose nitrate, but do not appear to be strong enough for very large-scale use.

Usually single enzymes are immobilized, but some work has been done on the co-immobilization of enzymes on to the surface of cells, or of two or more enzymes so as to form a multifunctional catalyst which should be capable of more complex conversions than is possible using a single enzyme. For instance, adenylate kinase and acetate kinase have been co-immobilized in polyacrylamide gel and used to regenerate ATP from AMP (and/or ADP) using acetylphosphate as the phosphorylating agent. The difficulty with this technique is to obtain operating conditions, such as pH and temperature, suitable for all the enzymes being used. The alternative is to use immobilized whole cells to carry out multi-step reactions.

7.3.2 Assessment of supports and methods

A great variety of natural or synthetic, organic or inorganic materials has been used as immobilization supports. These materials differ in size, shape, density and porosity, and can be used in the form of sheets, tubes, fibres, cylinders or (most popularly) spheres. The size of the support particles used is a crucial factor, as this helps determine the extent of internal diffusional restrictions on enzyme activity, the pressure drop generated in packed-bed columns, the velocity of fluid flow required to fluidize columns, and the power input needed to suspend particles in stirred reactors.

Unfortunately, most of the support materials and immobilization methods described in the scientific literature are not immediately suitable for large-scale use because of the cost or hazardous nature of the chemicals used, the poor mechanical properties of the supports, or because the methods are difficult to scale up or require highly skilled labour.

The immobilization method chosen should be simple, reproducible, mild, cheap, safe, versatile and easy to use on a large scale. Ideally, it is desirable if the immobilization method is specific for the enzyme of interest, so that simultaneous purification and immobilization could be performed. This has been shown to be possible using monoclonal antibodies to carry out the immobilization. No generally accepted 'best' method for immobilizing enzymes of cells is recognized: therefore it is very important to carry out a screening exercise early in an investigation whereby the available methods are compared, because the activity, stability and ease of use of the immobilized biocatalyst can vary enormously and unpredictably with the immobilization method used.

Many factors must be considered when choosing an immobilization method. These include the chemical nature and kinetic features of the reaction, the chemical and physical stability of the reactants and the biocatalyst, and the effect immobilization may have on the activity and stability of the biocatalyst, the yield and purity of product achieved as well as many other more specific effects.

In a typical immobilization experiment, the objective is to achieve as high an active enzyme loading per unit volume of immobilization support as possible. The following parameters are normally measured: the volume, enzyme activity and protein content or viable cell count of the enzyme or cells used; the weight, particle size distribution, porosity and the chemical and physical nature of the support used; and the activity and protein concentration of any biocatalyst remaining after immobilization has been completed. The operational stability (half-life) and productivity and resistance to microbial contamination are also measured. Note should be taken of any swelling, shrinking, aggregation or fragmentation of the support during immobilization or upon contact with substrate, of any leakage of the biocatalyst from the support, and also of the ease with which substrate can diffuse into the biocatalyst particles. Detailed guidelines concerning the characterization of immobilized biocatalysts have been issued by the EFB Working Party on Applied Biocatalysis (see *Enzyme Microb. Technol.* (1983) **5**, 304–307).

7.3.3 Effectiveness factors for immobilized enzymes

The combined effects of the factors which affect the intrinsic properties of the enzyme, together with any losses in activity incurred during immobilization, are expressed as *effectiveness factors*, which are also sometimes referred to as coupling efficiencies, the percentage retention of activity, or activity yields. The effectiveness factor represents the activity of the immobilized enzymes or cells divided by the activity of an equivalent quantity of free enzymes or cells assayed under the same conditions. Stationary effectiveness factors (n) are measured under initial rate conditions, and operational effectiveness factors (n^1) are measured under the conditions required in a production facility. Operational effectiveness factors are by far the most useful guide to the practical usefulness of an immobilized enzyme preparation. An effectiveness value of 1.0 indicates good reaction control, with no appreciable reduction in the activity of the enzymes or cells by immobilization or diffusional restrictions, etc. Values less than 1.0 give an indication of the amount of activity lost during immobilization plus the extent of diffusion and other limitations on enzyme activity. Thus, when using an expensive or scarce enzyme it is obviously advantageous to seek a high effectiveness value, whereas when using a cheap abundant enzyme, a high enzyme loading is usually more important so as to achieve a rapid conversion. It is possible, however, to obtain effectiveness factors higher than unity, for instance when cell division occurs inside immobilized cell preparations, or when inhibitors are selectively excluded from the support matrix and thus from the micro-environment of the enzyme.

7.3.4 The kinetics of enzymes in industrial use

An understanding of the kinetics of an enzyme-catalysed reaction is important, because it is a useful guide to understanding the reaction mechanism and thus optimizing the reaction, and also enables a degree of control to be exerted over the enzyme-catalysed reaction.

In 'classical' enzymology, initial rate determination of enzyme activities are

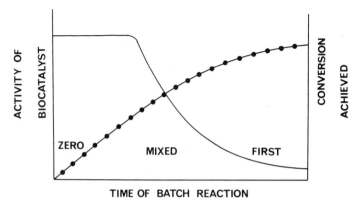

Figure 7.5 An illustration of the effect of the time for which a batch, enzyme catalysed, reaction (zero, mixed or first order) is allowed to proceed on the activity of the enzyme (———), the order of the reaction and the degree of conversion of substrate to product (●—●).

made over short time periods, in aqueous solutions and using dilute enzyme and substrate solutions. In industry we are generally interested in the complete conversion of concentrated substrate solutions using concentrated enzyme preparations, continuously for long time periods, often in non-homogeneous solutions containing immobilized enzymes or organic solvents. Furthermore, mass-transfer limitations, sometimes involving gases such as oxygen (see Chapter 4), may have so great an influence on the intrinsic kinetics of the enzyme as to make standard kinetic plots useless for practical purposes. In addition, in industrial applications, transient or pre-steady-state rather than steady-state behaviour is of little interest, except perhaps in attempts to predict response times and recovery times between start-up and shut-down periods resulting from cleaning and maintenance.

Because low average substrate concentrations (due to high substrate conversion with time) often prevail during industrial reactions, first- or mixed first- and zero-order kinetics apply, rather than zero-order Michaelis–Menten kinetics (Figure 7.5). The observed rate of reaction v is proportional to the prevailing substrate concentration S. Using the Michaelis–Menton equation:

$$v = \frac{V_{max}[S]}{K_m + [S]}$$

if $[S]$ is small, and because $V_{max} = k_2[E]$

$$v = \frac{V_{max}[S]}{K_m} = \frac{k_2[E][S]}{K_m} = \frac{-d[S]}{dt}.$$

Therefore, at a particular substrate concentration, it is the amount of active

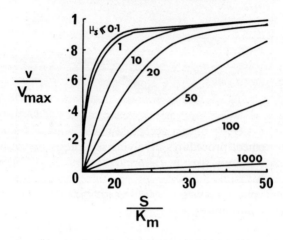

Figure 7.6 The effect of varying the bulk substrate concentration relative to K_m on the activity relative to V_{max} of an immobilized enzyme v/V_{max}) under different degrees of external diffusional restriction (μ_s from 0.1 up to 1000), and at various substrate concentrations (S/K_m). The external diffusional coefficient (μ_s) = $V_{max}/K_m hs$ where hs is the external transport coefficient. Redrawn from Engasser (1980).

enzyme present that determines the throughput of the conversion, and the time required to achieve a given product concentration is determined by the concentration of enzyme used. In a batch reactor, the highest rates of conversion (zero-order—see Figure 7.6) occur very soon after starting the reaction, when the substrate concentration is still high. The catalytic efficiency of the enzyme then decreases during the course of the reaction so that increasingly large concentrations of enzyme are required to achieve the same rate of reaction when progressively more highly converted substrates are formed. Because of the increased contact times between reactants and enzyme(s) required to achieve these high degrees of conversion, the possibility of chemical or biological side-product formation or further metabolism of the desired product is increased, such as the formation of colour and psicose during the production of high-fructose syrups by immobilized glucose isomerase. When multi-enzyme preparations or whole cells are used, very similar kinetic expressions results, because invariably one relatively inactive enzyme will rate-limit the overall reaction.

The kinetic behaviour of immobilized whole cells is complicated by the presence of additional barriers to substrate and product diffusion, the cell wall and membrane; and by the presence of active transport systems called permeases which transport some large molecules, such as sugars, into the cell. Most molecules, including O_2, move by diffusion, which is governed by Fick's law, but the active transport of nutrients is described by a Michaelis–Menten type of relationship, because it is based upon saturation kinetics behaviour of the permeases.

Because it is useful to maximize the degree of conversion of substrate into product, it is important to take into account the equilibrium constant K for the reaction being studied. If the equilibrium constant is very high, the conversion can be treated as an essentially irreversible reaction. If, however, the equilibrium constant is low, then the reaction is kinetically reversible and it may only be possible to obtain acceptably high extents of conversions by relatively inconvenient procedures involving separating the desired product from the reactor output stream, followed by recycling of the remaining substrate. Such reversible reactions can be utilized to great advantage. For instance the enzymic synthesis of the high-intensity sweetener Aspartame depends on using an immobilized metalloprotease, thermolysin, 'in reverse' to synthesise the the peptide bond between the aspartic acid and phenylalanine methyl ester components. If two or more reactants are involved in the reaction, they can be completely converted in a single pass through the reactor only if the reaction has a very high equilibrium constant, and if they are all supplied in the correct stoichiometric proportions. However, when one reactant is very much cheaper than the other(s), it may be advantageous to supply it in excess so as to maximize utilization of the more expensive reactant(s).

In heterogeneous catalysis, the rate of reaction is determined not simply by the prevailing pH, temperature, and substrate concentrations, but by the rates of ion transfer, heat transfer and mass transfer from the bulk substrate solution, through the support matrix to the immobilized biocatalyst. Diffusion of large molecules inside the immobilized enzyme support will obviously be limited by steric interactions with the matrix, and it is generally found that the activity of immobilized enzymes towards high-molecular weight substrates is lower when low-molecular-weight substrates are used.

The intrinsic kinetics of an enzyme are defined as being those of the soluble enzyme, and the kinetics of the immobilized biocatalyst in the absence of, and in the presence of, modifying factors are called the inherent and effective kinetics respectively.

7.3.5 *Factors which modify the intrinsic activity of enzymes*
Four main factors have been identified which modify the intrinsic catalytic properties of enzymes, either during or after immobilization. (i) Conformational effects are due to chemical modification of the enzyme protein during immobilization. These can have especially serious effects on the enzyme activity when amino-acid residues which form part of the active site, or which are important in maintaining the tertiary structure of the enzyme, are involved. (ii) Steric effects occur because often some of the enzyme

molecules are immobilized in a position relative to the support surface such that the active site is relatively inaccessible to substrate molecules. (iii) Microenvironmental effects occur because when immobilized the enzyme acts in a very different environment to that encountered in the bulk solvent. Partitioning effects are a common cause: for instance, hydrophilic substrates will be selectively attracted to the surface and pores of hydrophilic supports, whereas hydrophobic substrates will be repelled from hydrophilic supports but selectively attracted into hydrophobic supports. Similarly, positively charged substrates and protons will be attracted into negatively charged supports, giving a local high substrate concentration and a lower pH inside the support. Partitioning effects of this type can cause the K_m and/or the pH optimum of the immobilized enzyme to differ from the values obtained for the free enzyme. Thus for a negatively charged support, the K_m of the immobilized enzyme for a positively charged substrate will be decreased and the pH optimum shifted to a more alkaline pH, because the pH inside the support will in reality be lower than that in the medium outside. Similar but opposite effects will be observed when using positively charged molecules and positively charged supports. (iv) Lastly, diffusional restrictions occur because the substrate must diffuse to the immobilized enzyme before reaction can take place. Under these circumstances, simultaneous diffusion, as described by Fick's law, and enzymic reaction occur, so that concentration gradients of substrate and product are set up around the inside of the pellets, especially when a highly active enzyme is used.

Diffusional limitations on the activity of immobilized biocatalysts are of the greatest importance, and will be discussed separately as external and internal effects.

External diffusional limitations are due to the restricted rate of diffusion of substrate in the thin film of poorly mixed fluid surrounding each support particle, the Nernst–Planck layer. External diffusional restrictions can be decreased (but not prevented) by increasing the degree of agitation in well-mixed reactors or the flow rate in tubular reactors; or by using a more concentrated, or less viscous substrate (Figure 7.6). Note the use of μ_s, the external diffusion coefficient.

Internal diffusional limitations arise because of the small size and tortuosity of the pores in the support, which prevent the forced flow of fluid inside the pores of the pellets under normal operating pressures. A localized reduction in the measured diffusion coefficient compared to the value obtained in free solution results, such that

$$D_e = \frac{D\psi}{\tau}$$

where D is the diffusion coefficient measure in free solution, D_e is the effective diffusion coefficient measured inside the support particles, ψ is the porosity of the particles and τ is a tortuosity factor which represents the path-length which must be traversed by molecules diffusing between two points within a particle. In general, the effective diffusion coefficient is proportional to the water content of the support and inversely proportional to the molecular weight of the substrate.

Such internal diffusional restrictions have been partially circumvented by the development of pellicular immobilized enzymes, where a thin layer of active enzyme overlies a catalytically inactive core, designed to impart mechanical strength to the particle. Only relatively inactive immobilized enzyme activities can be obtained by this type of immobilized enzyme, but an additional advantage is that the pH in the vicinity of the enzyme can be controlled reasonably well.

High concentrations of enzyme can also contribute to internal diffusional restrictions such that the potential immobilized enzyme activity is not all expressed. This is because the enzyme molecules located in the outer portions of the support particle react at a rate greater than the rate at which substrate diffuses in, and so consume most or all of the substrate which enters the particles, so that the enzyme molecules located deeper in the particle 'see' little or no substrate. Thus even when low internal diffusional restrictions occur in the presence of a low concentration of immobilized biocatalyst, appreciable restrictions will be observed when higher concentrations are used, so that an optimum effective biocatalyst loading usually exists.

The significant variables which affect the extent of internal diffusional restrictions are quantitated by means of the Thiele modulus (ϕ), which has the form

$$\phi = L\lambda = L\sqrt{\frac{V_{max}}{K_m D_e}}$$

where L is half the thickness of the particle. When ϕ is large, the effectiveness factor for the immobilized biocatalyst is low, and when ϕ approaches zero the effectiveness factor tends to unity (Figure 7.7).

Internal diffusional restrictions can be recognized if the activity of an immobilized enzyme is increased when it is crushed, i.e. when the length of the diffusion pathway is reduced. Internal diffusional restrictions can be minimized by using a low-molecular-weight substrate, a high substrate concentration, a low biocatalyst concentration, and small, highly porous support particles in which the pores are as large, non-tortuous and interconnected as possible (Figure 7.8), or by immobilizing the enzyme only to the outside surface of the support. Thus there is a compromise between forming a very active immobilized preparation with a low effectiveness factor, and a less active preparation with a higher effectiveness factor. In fact, the

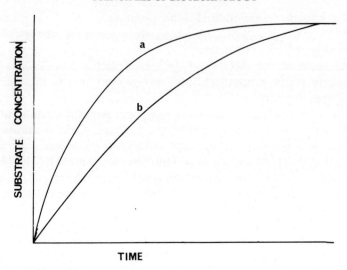

Figure 7.7 Measurements of the rates at which substrate diffuses into porous immobilized cell pellets packed in a column. Substrate solutions were pumped up the columns at (*a*) faster and (*b*) slower flow rates. The substrate concentration in the column eluate was measured until it reached the same concentration as in the substrate solution. Adapted and redrawn from Cheetham *et al.* (1979).

effectiveness factor will vary throughout the support particles, being high near the surface and lower nearer the centre of the particles.

Diffusional restrictions will also tend to underemphasize the effects of inhibitors, so that the combined effects of diffusional restrictions and chemical inhibition are less than the sum of their separate effects. Furthermore, an inhibitor present in the substrate will tend to inhibit the first enzyme molecules encountered, usually those located on the outside of the support and at the input end of a packed-bed column, and then progressively inhibit enzymes located deeper and deeper into the support particle and further along the column. Diffusional limitations on substrate and product transfer are also important in multiphase reactions, for instance where substrate molecules must pass from an organic solvent phase into an aqueous phase to react, and then the product must diffuse back into the aqueous phase.

7.3.6 *The stability of immobilized biocatalysts–diffusion artefacts*

The stability of an enzyme depends on the characteristic nature of the enzyme and the conditions under which it is used. The factors which stabilize and inactivate enzymes are not systematically understood. Individual enzymes vary very much in stability, and immobilization can influence the operational stability; both increases and decreases in stability having been observed experimentally. For instance, proteases are more resistant to autodigestion

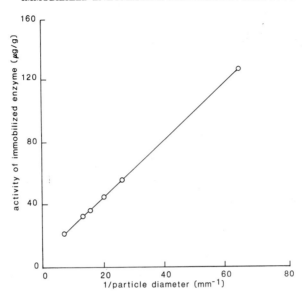

Figure 7.8 Immobilized esterolytic activity as a function of particle diameter for chymotrypsin immobilized to several size fractions of rock magnetite (Monroe et al., 1977).

when used in an immobilized form, and covalent linkage to a support can stabilize the tertiary structure (conformation) of enzymes due to the formation of covalent bonds that hold the enzyme in an active conformation. Some reports of the enhancement of operational stabilities by immobilization, may, however, be artefactual. This is because of diffusional restrictions, which allow only a fraction of the total immobilized enzyme to be active initially, the remainder acting as a reserve of fresh activity which only comes into action as some of the activity expressed initially is denatured (Figure 7.9). Thus a loss of activity with time is only observed after a period of time sufficient for this 'buffer' enzyme activity to be lost. In the absence of diffusional restrictions, activity decays exponentially with time, whereas when diffusional limitations are present activity decays linearly with time (Figure 7.10). A sudden loss of activity can also result from mechanical blockage or microbial contamination, which can often occur relatively suddenly.

7.4 Enzyme reactors

An enzyme reactor is the container in which a reaction, catalysed by free or immobilized enzymes or cells, takes place, together with associated sampling and monitoring devices. The reactor's role in a process is to produce a specified product in a given time from defined reactants at a minimum cost.

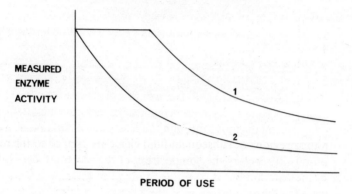

Figure 7.9 A diagrammatic representation of the operational stabilities of (1) immobilized and (2) free (soluble) enzyme preparations versus time. Note that the immobilized preparation is often more stable than the soluble enzyme and displays a period during which no enzyme activity appears to be lost.

Enzyme reactors differ from chemical reactors chiefly because they function at low temperatures and pressures, and differ from fermentations involving growing cells in not behaving in an autocatalytic fashion.

Reactors are classified according to whether their contents are homogeneous, in which only one phase is present, or heterogeneous, in which

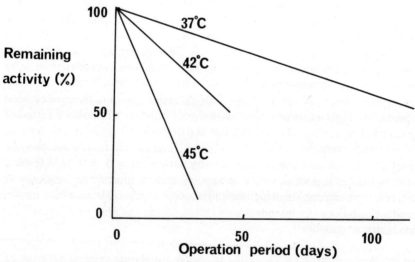

Figure 7.10 The operational stability of the aspartase activity of immobilized *E. coli* cells operated at three different temperatures (37°C, 42°C, 45°C). Note that the immobilized enzyme is less stable, but more active as the operating temperature is increased. Adapted and redrawn from Chibata et al. (1976a).

more than one phase is present; these are often a liquid substrate solution and a solid immobilized biocatalyst. Reactors can also be classified as to whether reaction takes place batchwise or continuously, whether they are open or closed (i.e. whether the catalyst passes out of the reactor during use or not), and according to the extent of mixing which takes place within the reactor, the idealized extreme types being 'perfectly mixed' and 'plug-flow' reactors. In the former type, molecules are maintained in a constant state of agitation, and in the latter, fluid elements move through the reactor as a 'fluid plug', without mixing with the previous or subsequent fluid elements applied to the reactor. In batch well-mixed reactors the composition of the reactants varies during the course of the reaction but is constant throughout the reactor, whereas in plug-flow reactors the composition of the reactants is time invariant and varies only along the length of the reactor. Thus in plug-flow reactors the effectiveness factor will be high near the input end and lower near the exit end, due to the decrease in substrate concentration as the reactants pass through the reactor, whereas in batch reactors the effectiveness factor is the same throughout the reactor but varies with the time of reaction, because the substrate concentration decreases with time. In practice, despite careful design and operation, the ideal types of mixing can only be approached, albeit fairly closely in many cases.

7.4.1 Types of biochemical reactor

A number of common enzyme reactor configurations exist (Figure 7.11). Additional features include, for instance, the incorporation of heat-exchangers inside columns to remove the heat of reaction. These simple types may be further complicated by the addition of recycle loops so as to facilitate complete conversion of substrate when high volumetric throughputs must be maintained.

Choice of the most appropriate type of reactor as well as their efficient design and operation is very important in making successful use of enzymes in industry. A multitude of factors should be considered. These include the need for pH and temperature control, the need to supply or remove gaseous reactants, the presence of particulate materials in the feedstock, the chemical and biological stabilities of the substrates and products, the frequency of catalyst replacement required, the presence of appreciable substrate and/or product inibition, the intended scale of operation and the likely uses of the product.

7.4.2 Assessment of the performance of biochemical reactors

The performance of a reactor is measured by its activity, stability and selectivity, and by the yield of product obtained, the degree of conversion of substrate to product which can be achieved, and the concentration of

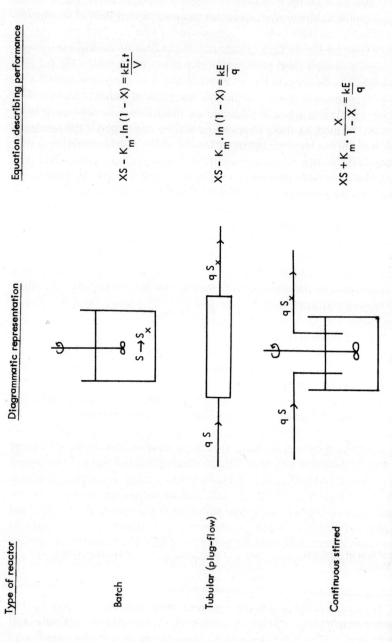

Figure 7.11 A diagrammatic and algebraic description of the common reactor types. Equations are derived for an enzyme acting on a single substrate in the absence of inhibition. X is the proportion of substrate converted, E is the total enzyme activity in the reactor, S and S_x are the initial and final substrate concentrations, q the flow rate of substrate into the reactor, t the time of operation and K_m the Michaelis–Menten constant.

reactants that can be used. These parameters should be measured under conditions which simulate the conditions expected in a full-scale production facility.

Activity is usually expressed volumetrically, as grams product formed per litre of reactor volume used per hour of operation (volumetric activity), or a similar term, as usually the aim is to achieve the required productivity as rapidly as possible and by using as small a reactor as possible. A high activity is especially desirable when a substrate or product is unstable, or if side-products are formed, as these phenomena will be minimized if the residence times of reactants in the reactor are as low as possible. Volumetric activities are influenced by the concentration of the immobilized biocatalyst, the amount of biocatalyst loaded on to the support material, and the proportion of this activity which remains enzymically active after immobilization.

The stability of a biocatalyst is usually expressed by means of a half-life value, i.e. the time required for the activity of the reactor to fall to half of its initial activity. It is obviously desirable for the activity to be as stable as possible, so that enzyme production and immobilization need only be carried out at infrequent intervals.

The selectivity of the biocatalyst is defined as the proportion of desired product formed divided by the total amount of products formed. Highly selective reactions are desirable because efficient use is made of the substrate supplied and high yields of the desired product favoured, especially if a high degree of conversion of substrate to product can be maintained. Furthermore, purification and isolation of the desired product is additionally favoured when high concentrations of reactants can be used, particularly when the enzyme(s) used are highly selective and the reaction is virtually irreversible. This is because the quantities of solvent (usually water) which must be removed before recovery of the product are greatly reduced. Thus the most important parameter describing the performance of a reactor is its productivity, that is, the volumetric activity multiplied by its operational stability, expressed for instance as kg of product formed per litre of reactor volume per year.

Monitoring of the material balance of the reactor is also important. Generally:

Rate of flow of reactants into the reactor	−	rate of removal of reactants by side-reactions or further reaction of the product	= rate of flow of product out of the reactor

For the normal batch reactor the first term is zero and the rate of accumulation of products or the rate of disappearance of the reactant equals the reaction rate. For a continuous tubular reactor or a continuous stirred reactor, no material should accumulate inside the reactor, and so the rate of removal of reactants by reaction is balanced by the difference between the inflow and outflow rates. Losses in material result from side-product formation, and the maintenance and growth requirements of cells, in the case of fermentations.

7.4.2.1 *Batch reactors.* Batch reactors are a versatile and traditional form of reactor which is especially useful when soluble enzymes are used or when small outputs or infrequent operation is required. Disadvantages include the tendency for conditions to change throughout the course of the reaction, and an inability to re-use the enzyme, or a tendency to lose enzyme during the recovery of the enzyme between batch reactions, although semi-continuous use is possible. These disadvantages are, however, often outweighed when a high-value product is formed. For instance, in the production of 6-aminopenicillanic acid from Penicillin V or G, batch stirred reactors containing immobilized penicillin acylase are used, because the acid formed during the reaction must be efficiently and rapidly neutralized by the continuous addition of alkalis. Today, however, more sophisticated reactors are commonplace. The two which have been most commonly used in industrial practice are the continuous stirred-tank reactor (CSTR) and the plug-flow column.

7.4.2.2 *Continuous stirred-tank reactors.* Stirred reactors consist of an agitated tank to which substrate is supplied at the same rate as the reactor contents are removed. They are well-mixed, so that the concentration of

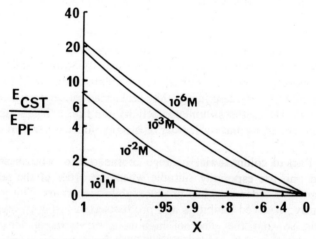

Figure 7.12 The variation in the efficiency of enzyme used in continuous stirred tank compared to packed-bed reactors (E_{CST}/E_{PF}) with the degree of conversion of substrate into product (X) at constant K_m ($= 10^{-3}$ M), and flow rates, but at various substrate concentrations (10^{-1}–10^{-6} M). Note the larger amount of enzyme required by the CSTR to achieve full conversion of substrate ($X = 1$). The CSTR has good pH control, good gas supply, but low concentration of immobilized enzyme. Redrawn from Lilly and Sharp (1968).

reactants is uniform throughout the vessel and the product stream leaving the reactor has the same composition as the mixture within the tank. CSTRs, or back-mix reactors as they are sometimes called, are cheap, versatile and especially suitable when liquid phase reactions are being carried out. Supply of gas, and pH and temperature control are easy, fresh catalyst can be easily added to the reactor, and substrates containing particulate materials can be tolerated without causing fouling. However, the relatively high power input required to give efficient agitation in a CSTR is clearly a disadvantage, and abrasion of the biocatalyst may occur. Also, only a comparatively low biocatalyst concentration can be maintained in a stirred reactor compared to the high concentrations possible in plug-flow columns, so that a given rate of product formation can be achieved using smaller column reactors. In practice a stirred reactor may nevertheless still be preferred, because large stirred tanks are easily and cheaply constructed. Because good mixing is achieved in stirred reactors, the reaction is likely to be less selective than in plug-flow reactors, and the biocatalyst will be more prone to product inhibition so that high degrees of conversion will be very difficult to obtain (Figure 7.12). Approximate plug-flow behaviour can be achieved by using a number of stirred tanks connected in series (Figure 7.13), although this is a comparatively complicated and expensive alternative. The best example of a commercial batch reaction is the de-acetylation of penicillins V or G by immobilized penicillin acylase, which must be carried out in a well-mixed reaction as alkalis must be added throughout the reaction to maintain a constant pH.

7.4.2.3 *Plug-flow column reactors.* Plug-flow (PFR) or tubular reactors consist of columns packed with the biocatalyst, usually in a particulate immobilized form. Plug-flow columns are kinetically more efficient than stirred reactors and require less enzyme when compared on a volumetric activity basis (Figure 7.12), and can be relatively simple and easy to operate and automate. They are best used continuously and on a large scale so as to minimize labour costs and overheads, and to facilitate control, resulting in more reproducible product quality than would be obtained by batch processes. Packed columns allow a high concentration of biocatalyst to be maintained and are especially suitable when the order of the reaction is greater than zero-order, or when product inhibition occurs. This is because endogenously produced inhibitors and products are constantly swept out of the reactor, so that the effect of high concentrations of substrate are minimized (Figure 7.14). However plug-flow reactors are more prone to substrate inhibition than CSTRs, as in the former, the substrate concentration decreases from the input end of the reactor to the output end, while in the CSTR uniform mixing ensures that the enzyme is exposed to relatively lower average substrate and product concentrations.

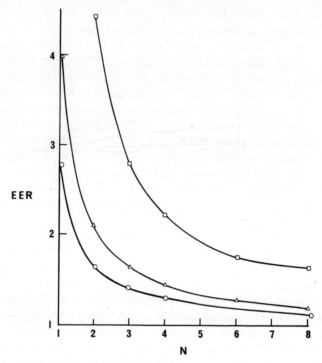

Figure 7.13 The effect of the number of stages (N) of a multi-stage continuous stirred reactor on the excess amount of enzyme (penicillin acylase) activity (EER) required to achieve various degrees of conversion of substrate (X), compared with the activity required by a batch reactor with no down-time between runs. ○—○, △—△ and □—□ represent X values of 0.99, 0.95 and 0.90 respectively. Redrawn from Carleysmith and Lilly (1979).

Only a relatively low power input is required by tubular reactors, but it is very difficult to maintain constant pH and temperature or to supply gaseous reactants, although these operations can be carried out in recycle loops. Only a limited physical access to the reactor is allowed, so that it is difficult to add fresh biocatalyst during reaction. Also, packed-bed columns can be prone to self-compression and to fouling by particulate materials in the substrate stream, and are more difficult to operate hygienically than batch reactors. Column reactors are conventionally used on a commercial scale for glucose isomerization, production of amino acids and amino acid resolution processes.

7.4.2.4 *Fluidized bed reactors.* In fluidized beds, the biocatalyst particles are maintained in suspension relative to each other by the upward passage of substrate or gas at high flow rates. Temperature, pH and gas supply are easily controlled using fluidized beds, and substrates containing solid particles can

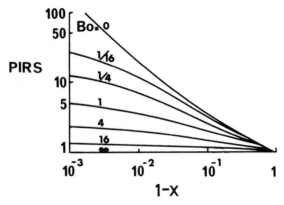

Figure 7.14 An illustration of the proportional increase in reactor size (PIRS) required to compensate for deviations from ideal plug-flow behaviour over a range of degrees of conversion of substrate $(1-X)$ and a S/K_m of 0.5. The degree of back mixing increases with the Bodenstein number (Bo), a measure of the extent of back mixing which equals infinity under plug-flow conditions and zero in perfectly mixed conditions. Redrawn from Kobayashi and Moo-Young (1971).

easily be dealt with. However, because of the large void volume of the fluidized bed, high biocatalyst concentrations cannot be achieved and the high flow rates of substrate solutions that are required to fluidize the biocatalyst particles may cause wash-out of the catalyst and allow only a low degree of reaction to take place. Also, because the reactor is well-mixed, high degrees of conversion are only achieved efficiently by using a high concentration of enzyme (Figure 7.14). Wash-out and complete conversion of substrate to product can be avoided by recycling the substrate, by using tapered fluidized beds or by using several fluidized beds in series. In the Kyowa–Hakko process for the production of non-potable alcohol using immobilized yeast cells, the column is partially fluidized by the CO_2 evolved by the cells.

7.4.2.5 *Ultrafiltration reactors.* Ultrafiltration reactors are well-mixed reactors which rely on a selective membrane to separate low-molecular-weight products from higher-molecular-weight substrates. They are therefore most useful for carrying out depolymerization reactions, especially when using soluble enzymes, so as to ensure good contact with macromolecular substrates. The main disadvantage associated with the use of ultrafiltration reactors are the small sizes of reactor available, and concentration polarization, that is blockage of the pores in the membrane by solid, fat or colloid particles present in the substrate. Similar selective membranes have also been used in hollow-fibre reactors, which can be used in a plug-flow fashion with substrate rapidly recirculated through the reactors, but with the enzyme or cells located inside the fibre.

7.4.2.6 *Electrochemical reactors.*
These are a specialized form of reactor in which redox enzymes are used, in combination with electrodes, as analytical devices, fuel cells to produce electricity, or to achieve chemical conversions using electrical energy. Difficulties arise in obtaining good coupling between the various redox components and with the electrodes, which need to be large where high electron flux, essential for efficient conversion, is required.

7.4.2.7 *Two-phase reactors.*
Two-phase reactors are a recent development, using either an aqueous phase and a water-immiscible organic solvent, or two partially miscible aqueous phase, such as PEG (polyethylene glycol) and dextran solutions. Such systems have several advantages: for instance, reactants which are insoluble in pure water can be used, and product can be recovered following partitioning from one phase, where the reaction takes place, into the other phase, in which it is more soluble. It appears that the performance of biocatalysts in two-phase reaction systems is best described by equations containing the term log P (the partition coefficiency of the reactants between the two phases).

7.4.3 *Practical enzyme reactor kinetics*

Enzyme reactor 'kinetics' are a development of the expressions derived for immobilized biocatalysts, taking account of the effects of the flow rate of substrate through the reactor, the degree of conversion of substrate to product, the extent of mixing of fluid in the reactor, and the influence of inhibitors. The overall effect of these parameters is to minimize the size of reactor or the concentration of enzyme required to achieve a given productivity.

Michaelis–Menten kinetics can be used as a basis for a kinetic description of batch reactors, but the equation is best used integrated with respect to time to give the total conversion of substrate achieved.

For an unihibited, irreversible reaction employing a single enzyme in a batch reactor under isothermal conditions, the equation describing performance is

$$XS - K_m \ln(1-X) = \frac{kE \cdot t}{V}$$

where kE is the maximum activity of the total enzyme in the reactor in moles of substrate converted per second, V is the working volume of the reactor, and t is the time of operation in seconds (see Figure 7.11).

Similarly, for a tubular or plug-flow reactor (PFR):

$$XS - K_m \cdot \ln(1-X) = kE/q = Ekr/V$$

where q is the volumetric rate of supply of substrate in s^{-1}, r is the average residence time of substrate solution in the reactor in seconds, and X is the degree of conversion, i.e. $(S - S_x)/S$, where S is the original substrate concentration and S_x the concentration remaining after a period t of reaction.

For a well-mixed continuous-flow reactor (CSTR):

$$XS + K_m \cdot \frac{X}{(1-X)} = kE/q = kEr/V$$

(see Figure 7.11).

Elaborations of these three basic equations have been developed to describe the effects of substrate and product inhibition.

An important experimental parameter is the term S/K_m. At low levels of S/K_m where a high degree of conversion has been obtained, or a low substrate concentration is used, the reaction rate is essentially first order. Here the time of reaction to give a particular degree of conversion is directly proportional to S/K_m. However, at high values of S/K_m where the reaction is essentially zero order, the time taken to reach a given degree of conversion is independent of S/K_m. In industrially practicable processes, high degrees of conversion of substrate to product are usually required and so the time required for a comparatively small increase in conversion is critically dependent on the prevailing value of S/K_m.

In continuous-flow packed columns or stirred reactors there is an important relationship between the flow rate through the reactor (q) and the degree of conversion (Figure 7.15). When a high S/K_m is used and low degrees of conversion are being achieved, the behaviour of both reactors is almost identical. However, at low values of S/K_m, i.e. when the order of the reaction is

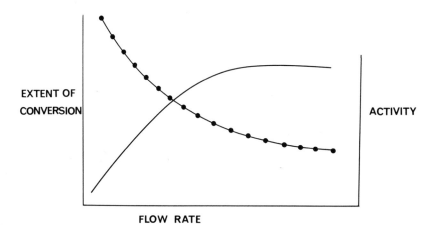

Figure 7.15 The effect of the flow rate of substrate through a packed bed column on the degree of conversion of substrate of products achieved (●—●) and the activity of the immobilized biocatalyst (———).

greater than zero, the performance of tubular reactors becomes even more favourable when compared to that of well-mixed reactors, as high degrees of conversion can be obtained at much higher flow-rates in tubular (plug-flow) reactors rather than in well-mixed reactors.

It is the quantity of active enzyme contained in each reactor that determines the extent of the reaction achieved, i.e. the percentage of substrate conversion achieved. Reactors are compared on the basis of the volumetric activity, for example, grams specification grade product per litre reactor volume. Thus, in Figure 7.15, the amount of enzyme required to give a range of conversions (zero to 100% shown on a scale of 0–1) in a continuous stirred reactor, compared with the amount of enzyme required in a packed column reactor, are plotted as ratios for various substrate concentrations. The E_{CST}/E_{PF} ratio at extreme values (at $X = 1$ in Figure 7.13) is identical to the relative sizes of plug flow and CST reactor required. In Figure 7.12, this is shown for zero-order ($S = 10^{-1}$ M) and first order ($S = 10^{-6}$ M) reactions, assuming no change in the volume of the reactors, so that E_{CST}/E_{PF} reflects the relative sizes of reactor needed to achieve the same productivity.

For batch reactors, the amount of enzyme required will be higher than for a continuous reactor because of the 'down-time' lost between batches. Thus, if α is the proportion of the continuous reactor operation time that the batch reactor is in operation, then the amount of enzyme required to give an overall productivity equal to that of the plug-flow reactor (E) will be E (plug-flow) $\times 100/\alpha$.

7.4.4 The effect of non-ideal flow on biochemical reactor performance

Non-ideal flow is caused by back-mixing, convection, or the creation of stagnant regions of fluid in the reactor, for instance by channelling. Back-mixing of reactants in packed-bed reactors arises because the presence of solid particles in the column causes elements of the flowing fluid to mix, and because the different fluid elements have different residence times in the reactor and so are in contact with the enzyme for different periods, giving an opportunity for differing extents of reaction to take place. The extent of non-ideal flow is measured by means of the residence time distribution of pulses of liquid dye as it leaves the reactor.

In stirred reactors, better mixing can only be achieved by increasing the stirrer speed, by decreasing the viscosity of the substrate or the concentration of biocatalyst particles, or by baffling the reactor more efficiently. In packed-bed reactors, back-mixing, channelling and temperature gradients within the reactor are the most important causes of deviations from plug-flow behaviour and are expressed as the increase in reactor size needed to achieve the required productivity (Figure 7.14). When S/K_m is large, the reaction is zero-order and non-ideal flow has no effect on the output of the reactor. However, when S/K_m is very small, i.e. when a high percentage conversion has been attained, back-mixing becomes very important and the volume of the reactor needed to achieve the required productivity increases, particularly in stirred reactors. Plug-flow is encouraged by the use of even-sized, smooth, spherical, evenly-packed support particles and by the absence of accumulated solids or gases in the column.

Back-mixing affects the yield of a reactor in several ways. Firstly, as described above, some substrate becomes mixed with the product stream leaving the reactor and so is no longer available to form product (Figure 7.14). Secondly, back-mixing decreases the substrate concentration, lowering the rate of reaction, and increasing the product concentration at which reaction is taking place. Thus non-ideal mixing also causes a marked decrease in the selectivity of the reaction because the residence time of some product molecules will be much greater than average, giving greater opportunity for side-product formation and loss of the desired product by further metabolism.

7.4.5 *The stability of biochemical reactors*

The useful lifetimes of biocatalysts are usually much shorter than those of chemical catalysts. The operational stability of an immobilized biocatalyst is affected by a number of factors. These include irreversible inhibition by substances present in the substrate or produced endogenously, denaturation by pH, temperature, ionic-strength shocks, organic solvents, shear forces, microbial attack, fouling of the column, leakage of cells or enzyme from the support, dissolution or fragmentation of the support, and poor *enzyme*–substrate contact, or self-digestion in the case of proteases. A good example is the inactivation of immobilized glucose isomerase by oxidation; it is thus normal practice to use de-aerated substrate.

Loss of enzyme activity can be regarded as a special type of enzyme inhibition, which may occur homogeneously throughout the particle at the same time, or selectively, starting at the surface of the particle and moving inwards. Decay increases with the amount of denaturant present and is a first-order process with respect to time (Figure 7.10). However, when the reaction is diffusion-controlled, as is usually the case, decay of activity is linear with time until the biocatalyst activity has been reduced to a point where the reaction is no longer diffusion-controlled. When the shape of the decay curve is 'concave', decay of activity can be caused by leakage of enzyme from the reactor during use, whereas 'convex' decay can be caused by cumulative inhibition or denaturation.

For first-order decay

$$\ln \frac{N_0}{N} = \lambda t$$

where N_0 is the original activity of the reactor, N is the activity remaining after time t, and λ is the decay constant. The stability of the reactor is usually described in terms of a half-life, which is the time required for half of the original activity to be lost:

$$\text{half-life} = 0.693/\lambda$$

or, alternatively,

$$\text{half-life} = \frac{\ln 2}{\ln(N_0/N)}$$

In practice, reactors are operated for a period after which it is calculated that it becomes more cost-effective to replace or regenerate the catalyst rather than to continue to use the original enzyme preparation. When complex metabolic pathways or whole cell activities are used, activity still decays in a similar fashion to when a single enzyme is being used, presumably because the stability of one relatively labile enzyme limits the overall rate of the system. As well as the operational stability, the storage stability of an immobilized biocatalyst is important because this property dictates the operator's ability to transport and store the immobilized cells or enzyme, or to use the immobilized biocatalyst intermittently.

7.4.5.1 *Regeneration of biocatalyst activities.* Regeneration of the activity of an immobilized enzyme reactor is not possible at present, except for a few enzymes which undergo reversible denaturation, other than by desorbing the exhausted enzyme and then immobilizing fresh enzyme. This regime is difficult to carry out *in situ*, unless the enzyme is immobilized by adsorption as in the case of immobilized amino acid acylase. Immobilized biocatalysts are therefore routinely disposed of at the end of their useful life and replaced with

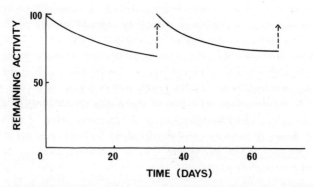

Figure 7.16 The stability and regeneration of aminoacylase immobilized on DEAE-Sephadex and used in a packed-bed column. Regeneration (denoted by ↑) was achieved by desorbing the exhausted enzyme and adsorbing fresh active enzyme. Redrawn from Chibata *et al.* (1976*b*).

fresh immobilized enzymes or cells (Figure 7.16). Regeneration of the activity of immobilized viable cells can, however, sometimes be carried out, either by re-inducing enzyme(s), or by supplying growth medium and allowing cell growth to take place *in situ*. Unfortunately, cell growth may lead to leakage of cells from the support particles and increased diffusional restrictions inside the pellets, firstly because the new cells may block pores in the support, and secondly because the fresh cells are often not evenly distributed throughout the support.

7.4.5.2 *Maintenance of a constant productivity and product quality from biocatalyst reactors*. Maintenance of a constant rate of output of product from a reactor throughout the useful lifetime of the reactor is important so as to match the output from the reactor to the throughputs through the equipment required for purification and isolation of the product; and to help maintain a constant output and quality of product from the plant. Constant productivity can be achieved by adding fresh biocatalyst during operation, but this procedure is not always operationally convenient, or even possible when packed-bed reactors are used. Alternatively, the flow rate through the reactor can be continuously or intermittently decreased so as to maintain the degree of conversion constant, although the quantity of product produced per unit time will fall during the period of operation. When this mode of operation is used it is important to express results as the flow rate required to give the desired product quality (degree of conversion) versus time, rather than the activity or percentage conversion achieved versus time. Another method of maintaining constant product quality is to increase the temperature of operation as the reaction proceeds so that the loss in enzyme activity with time is offset by the increased enzyme activity obtained at the higher temperatures. Changes in the flow rate or temperature can be easily and accurately controlled and recorded by means of microprocessor devices. The latter approach, manipulating temperature, has been used in the industrial production of malic acid using the fumarase activity of immobilized *Brevibacterium ammoniagenes* cells. However, the activities of biocatalysts are invariably much less stable when higher temperatures are used (Figure 7.10) and so the overall productivity of the reactor is reduced.

A number of columns of different ages, out of phase with each other, are often used in combination with one of the above procedures (Figure 7.17). Thus the amount of the immobilized biocatalyst used does not change with time, as exhausted columns are continually being replaced with fresh columns. The greater the number of reactors that are used, the smaller the variation in product concentration that will result. In some cases where an expensive support is used, the exhausted enzyme is removed by pyrolysis and the regenerated support reused. In others, such as when using glucose isomerase the exhausted immobilized cells are used for cattle feed.

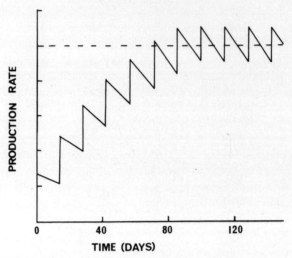

Figure 7.17 The achievement of a steady rate of production of hydrolysed cheese whey (— — —) using an assembly of 7 columns of lactase immobilized on porous silica, each with a 50 day half-life and used continuously for 100 days. Redrawn from Pitcher (1978).

7.4.6 *Physical problems associated with the use of immobilized biocatalysts in biochemical reactors.*

The chief mechanical problems associated with the use of enzyme reactors are abrasion in stirred reactors, and compression and fouling in packed columns. Abrasion of biocatalyst particles in CSTRs or fluidized beds increases with the shear rate and with the fraction of the reactor volume occupied by the particles, and decreases as the viscosity of the suspending fluid and the strength of the support particles are increased.

Compression of biocatalyst particles in packed-bed reactors occurs when excessive pressure-drops through the column are generated due to friction between the fluid and the support particles or due to partial blockage of the column bed by particulate materials. Compaction of the packed bed is especially likely when small or irregularly shaped or packed particles are used in large, tall columns for long periods and limit the practical size of industrial immobilized biocatalyst columns, for example for glucose isomerase. The pressures generated cause deformation or fracture of the pellets and result in a gradual decrease in the void volume between the pellets and eventually complete blockage of the column (Figure 7.18). Compaction can be minimized by using relatively large, incompressible, smooth, spherical and evenly packed support particles, or sectionalized columns, by fluidizing the bed at intervals or by decreasing the height to diameter ratio of the column. Creep phenomena are important determinants of the rate of compaction, creep being the very slow but continuous deformation of materials that are

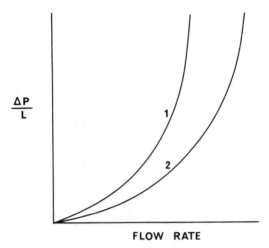

Figure 7.18 The influence of the flow-rate through a packed bed reactor on the pressure drop per meter length of column ($\Delta P/L$), developed across the bed of the column. Note that excessive flow rates will cause excessive pressures and eventual mechanical failure to develop. Curves representing the performance of columns containing (1) small and (2) large particles are depicted.

subject to relatively low stresses for long periods, like the deformation of fluids of extremely high viscosity such as glass.

Fouling causes a loss in the activity of immobilized biocatalysts by the deposition of solid or colloidal materials which were originally suspended in the substrate, either around, or in, the pores of the support, so preventing access of substrate to the enzyme. In stirred reactors fouling of inlet or exit filters is the main problem, whereas in packed-bed reactors fouling of the void spaces between particles and of the pores of the particles can occur. The extent of fouling depends on the presence, amount, size distribution and chemical and physical nature of the fouling materials and on the location of the biocatalyst either in, or on, the support material, and is accentuated when concentrated viscous substrates are used. Fouling can be prevented by clarification of the substrate by filtration or centrifugation, for instance prior to the treatment of whey by immobilized lactase columns. These, however, are expensive and inconvenient procedures, and cannot be used when the fouling materials are an essential part of the product, as in the case of milk. Fouling of columns may be prevented by intermittent repacking, fluidization or backflushing of the column bed.

7.4.7 *The purification and recovery of the products of biochemical reactors (downstream processing)*

Methods for the purification and isolation of the chemicals produced by biochemical reactors, commonly referred to as downstream processing, are

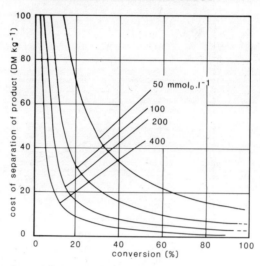

Figure 7.19 Cost of separation per unit weight of product as a function of conversion at different starting substrate concentrations. From Wandrey and Flaschel (1979).

essential if rather unspectacular component operations in any biotechnological process. Often they consume the majority of the capital and operating costs of a project, particularly when rigorous purities are demanded of the final product by customers and legislative authorities (Figure 7.19).

Usually the desired product(s) are present in low concentrations; for instance, antibiotics are present at levels of only a few gl^{-1} in fermentation broths, and vitamin B_{12} occurs at concentrations much smaller than this. Also, the desired product is invariably contaminated by a variety of other enzymic and chemically derived materials which may vary greatly in molecular weight, solubility and chemical and physical nature. Therefore purification and concentration of the desired product is a difficult operation, particularly if it must first be liberated from inside cells, or from association with cellular material. Purification and separation is of course facilitated if side-products or the desired products are produced as a different phase. For instance, the CO_2 produced by fermenting yeast cells is easily removed from the desired product, ethanol, because it is a gas rather than a liquid. Similarly, formation of solid products from liquid substrates, or vice versa, are desirable, such as the precipitation of aspartame as an addition compound with excess of the aspartic acid substrate, which leads to simultaneous recovery and purification.

There is only a limited number of separation techniques available for treating biological materials on any significant scale. This is probably partially because insufficient systematic work has been done on developing

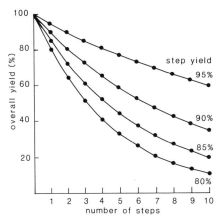

Figure 7.20 The effects of the number of operational steps and the step yield on the overall process yield. From Fish and Lilly (1984). © 1984, *Bio/Technology*. Used by permission.

methods that are especially suitable for biological materials, so that usually chemical engineering methodology and equipment, tried and tested in chemical processes but relatively unproven when used with biological materials, is adapted. A second tendency is to assume that for relatively high-value, low-bulk products, scaled-up laboratory methods such as affinity chromatography and gel filtration will prove adequate despite their small scale of operation, inconveniences, inefficiencies and high cost. In the future, more efficient, rapid, cheaper, larger-scale and preferably biospecific methods which have been designed to give high yields of product with the required purity and physical form, with the minimum cost and effort, are required. A particular requirement is to reduce the number of operations involved in processes (Figure 7.20).

The methods in common use include the various types of solid–liquid separation such as flotation, flocculation, centrifugation and filtration, and precipitation by pH changes, solvents such as acetone, or by salts such as ammonium sulphate. Other techniques include solvent extraction, crystallization, distillation, and the various filtration methods using filter presses, rotary vacuum filters or micro- or ultrafiltration devices, batch and column ion-exchange chromatography and the various methods of drying such as freeze, spray, fluidized-bed, tray or drum drying. Many of these methods, such as ion-exchange chromatography and solvent extraction, can be used so as to simultaneously purify and concentrate the desired material. Problems include the instability of biological materials to organic solvents and the requirement for hygienic operation, which is especially important when microbial contamination can occur and food or pharmaceutical use of the product is intended. More specifically, harvesting microbial cells when the cells are small and/or the culture medium is viscous is difficult, and there are problems in scaling-up chromatography columns, including the major problem of fouling by particulate materials. The reader is referred to the applications of these techniques to the isolation and purification of enzymes, discussed in Chapter 6.

A new approach, often termed 'extractive bioprocessing', involves the integration of the biocatalyst reaction stage with isolation of the product. This not only has the advantage of reducing the number of steps in the process, but also allows product inhibition to be more easily overcome and enables the equilibrium point of a reaction to, in effect, be shifted towards product formation. Product removal can be achieved by adsorption to ion-exchange resins, passage across an ultra-filtration membrane, or (if volatile) by removal from the exit gases from the reactor.

7.5 Conclusions

The field of enzyme engineering, involving the industrial use of biocatalysts, is still young, and relatively few immobilized biocatalyst processes are operated on a commercial scale so far. Many major problems need to be overcome. These include the stabilization of biocatalysts, so as to prolong their operational lifetimes, the ability to use water-insoluble and water-immiscible substrates such as steroids, hydrocarbons and gases, the use of substrates containing undissolved solid particles, the ability to regenerate cofactors *in situ* so that synthetic reactions can be performed, and the use of co-immobilized cells, enzymes, organelles, etc., so that complex multistep reactions can be carried out. Also, efficient methods for harvesting microbial cells, and cheaper, more selective and efficient methods of isolating and purifying the biochemicals produced by enzymic catalysis need to be developed. Most importantly, an increased range of biocatalysts with different activities and characteristics such as thermostability are required in cheap readily available forms. Protein engineering and cloning techniques promise to make modifications to enzyme properties possible, and allow more efficient production of enzymes from safe microbial sources. Improved down-stream processing techniques, such as membrane separation methods, could improve the isolation and purification of products and so decrease production costs. For instance many new uses could be found for a cheap, readily available lignase; also, no cephalosporin C acylase has been found which could be used to carry out analogous processes to those based on the use of penicillin acylase for the production of semi-synthetic antibiotics.

The following general points facilitate the successful application of biological catalysts in industry. Secure supplies of cheap, abundant (but unfortunately often relatively impure) substrates are preferred, and should also be used for preliminary experimentation. High substrate concentrations are preferred, so as to decrease the volumes of fluid which must be processed and the volumes of solvent, usually water, which must be removed during recovery of the product. The maximum possible degree of conversion of substrate into product should be achieved throughout the lifetime of the reactor, so as to minimize wastage of substrate and to facilitate isolation of the products. Thus, the use of biocatalysts in which the position of equilibrium (K) greatly favours the formation of products is very important. Highly active enzyme preparations are desirable, so as to decrease the size of reactor needed to achieve the required productivity. The purification and isolation of the product is simplified if the required products are formed in a different phase from the other reactants or side-products, or are easier to purify by some other simple technique such as precipitation. Similarly, maintenance of sterile or hygienic conditions in the reactor is facilitated if products such as antibiotics, organic acids or alcohols, which have antimicrobial activities, are formed.

In this chapter, the major concepts and practices associated with the use of immobilized biocatalysts are outlined, but only those immobilized cell/enzyme process which are in industrial use, or which have potential large-scale application, have been discussed. These processes chiefly involve the use of biocatalysts carrying out relatively simple hydrolysis or isomerization reactions mediated by a single enzyme.

Finally, the interdependence of all the unit operations which go to make up a process should be recognized and the overall importance and scientific and economic viability of the entire process, rather than any individual step or operation, however scientifically novel it may be, should be given priority. Only then can the diverse catalytic potential of enzymes be realized on an industrial scale, not only to produce high-value, low-bulk products such as pharmaceuticals, but also to produce low- or high-value or bulk products, such as fuels and solvents, from renewable natural resources.

References

A more comprehensive and detailed account of immobilization and reactor studies is given in *Methods in Enzymology*, vols. 44 and 136 (ed. K. Mosbach) Academic Press (1976 and 1987).
Up-to-date papers and reviews on all aspects of biotechnology, including immobilized enzymes and cells, can be found in the series *Advances in Biochemical Engineering* (Springer Verlag), *Topics in Enzyme and Fermentation Biotechnology*, vols. 1–10 (1977–85). (Ellis Horwood) and *Enzyme Engineering* (Plenum Press), and in the journals *Biotechnology and Bioengineering*, *Enzyme and Microbial Technology*, *European Journal of Applied Microbiology and Biotechnology and Bio/Technology*.

Carleysmith, S. W. and Lilly, M. D. (1979) *Biotechnol. Bioeng.* **21**, 1057–1073.
Cheetham, P. S. J., Blunt, K. W. and Bucke, C. (1979) *Biotechnol. Bioeng.* **21**, 2155–2168.
Cheetham, P. S. J., Imber, C. E. and Isherwood, J. (1982) *Nature (London)* **299**, 628–631.

Chibata, I., Tosa, T. and Sato, T. (1976a). In *Methods in Enzymology*, vol. 44, ed. K. Mosbach, 739–746.
Chibata, I., Tosa, T., Sato, T. and Mori, T. (1976b) *Ibid.*, 746–759.
Chibata, I. and Tosa, T. (1977) *Adv. Appl. Microbiol* **22**, 1–25.
Fish, N. M. and Lilly, M. D. (1984) *Bio/Technology* **2**, 623–627.
Gotfredsen, S. E., Ingvorsen, K., Yde, B. and Andresen, O. In *Proc. Symp. Biocatalysis in Organic Synthesis*, ed. H. Tramper *et al.*, Elsevier, Amsterdam.
Kobayashi, T. and Moo-Young, M. (1971) *Biotechnol. Bioeng.* **13**, 893–910.
Monroe, P., Dunnill, P. A. and Lilly, M. D. (1977) *Biotechnol. Bioeng.* **21**, 1629–1638.
Lilly, M. D. and Sharp, A. K. (1968) *Chem. Eng.* (London) **215**, CE12.
Pitcher, W. H. (1978) In *Enzyme Engineering* 4, eds. G. G. Broun, G. Manecke and L. B. Wingard, Plenum, Oxford, 67–76.
Poulsen, P. B. and Zittan, L. (1976) In *Methods in Enzymology*, vol. 44, ed. K. Mosbach, 809–821.
Rovito, B. J. and Kittrel, J. R. (1973) *Biotechnol. Bioeng.* **15**, 143–161.
Wandrey, C. and Flaschel, E. (1979) *Adv. Biochem. Eng.* **12**, 147–218.

Index

acetic acid in vinegar production 18–19
Acetobacter, ethanol oxidation by 19
Acinetobacter in microbial community 26
adsorption, immobilization method for 170–1
aeration number (N_a)
  formula for 84
  in fermenters 124
aerobic process in vinegar manufacture 89
affinity chromatography in enzyme purification 158, 161–3
aflatoxin from *Aspergillus flavus* 137
agitation of cells with abrasives 145–6
agitator (impeller) in fermenters 89, 104–8
agitator power dissipation in fermenters 106–7
air-lift fermenter
  in citric acid production 17–18
  engineering features of 107–10
  for SCP production 131–3
air sparging into fermenters 88
air sterilization for fermenters 101
alcoholic beverages 10–13
ale fermentation by *Saccharomyces cerevisiae* 12–13
alkali for cell disruption 148–9
alkylating agents, mutagenic 49–51
amino acids
  fermentation principles of production of 16–17
  production of 13–17
  submerged-culture production of 13
  therapeutic use of 13
aminoacylase process 172
aminoacylase, use of in immobilized form 137
ammonia, assimilation of by *Methylophilus methylotropus* 27–30
ammonium sulphate, enzyme precipitation by 152–3
amyloglucosidase, uses of 137, 141
α-amylase activity
  in saccharification 12
  uses of 137, 139–42
ancillary processes in fermentation industries 96–7
aneuploidy 51–4
animal cell culture *see* cell culture
animal cells as enzyme source 145
antibiotic-committed enzymes 8–9
antibodies, monoclonal 3, 159–63
anti-foams in fermenters 105
apparent stabilization, immobilized biocatalysts and 180–1

Arthrobacter, amino acid production by 14
aseptic conditions in fermentation 86–7
aseptic operation, requirements for 98, 99
aspect ratio in fermenters 34
assimilation, methane 26
Aspergillus flavus, aflatoxin from 137
A. foetidus, citric acid production by 89
A. nidulans, genetics of 60–1
A. niger
  citric acid production from 17
  glucose oxidase in 18
  surface culture of 88
A. oryzae 11
A. terreus, surface culture of in itaconic acid production 88
attenuator, transcription 45
auxotrophic bacteria, amino acid production by 15

Bacillus ammoniagenes, production of 5′-IMP by 24
B. brevis in biosynthesis of gramicidin S 8–9
Bacillus spp. 11
  transformation in 53–5
B. subtilis 8–9
  vectors for 64–5
back-mixing in enzyme reactors 192–3
bacterial chromosome 49–51
bacterial plasmids 49–51
bacteriophages as vectors 64–5
baffles in fermenters 104–6
baker's yeast 93–4
Bam H1 restriction endonuclease 62–3
barley malt 11–12
base substitution in mutation 49–51
batch culture 5–6
batch and continuous processes, comparison of 91–2
batch operation, fermentation engineering aspects of 90
batch reactors for enzymes 184–6
beer, enzymes used in chill haze prevention in 141–2
beer production, fermenters in 128–30
beverages, alcoholic 12
Bingham plastic fluid behaviour 119–21
biocatalyst stability, half-life value for 185
biocatalysts, immobilized 168–81
biocidal fluids 115–18
biological catalysts, use of in industry 164–5
biological waste-water treatment 130
biomass ($\bar{x}$)
  concentration of in chemostat 22–5

biomass ($\bar{x}$)—continued
  production rate of 6–7
  variation of with dilution rate 23–5
biomass accretion, rate of in continuous culture 23–5
biotechnology, definition of 1
Bodenstein number of 187–8
bottom fermenting yeast 12
pBR322 vector, structure of 64
brewing vessels 128–30
  fermentation in 88–9
  wort in 12
Brevibacterium flavum 14
Brevibacterium sp., amino acid production by 14
British Co-ordinating Committee on Biotechnology (BCCB) 3

Candida lipolytica
  citric acid production by 17–18
  growth of on *n*-paraffins
C. utilis in Symba process 27
carbon dioxide, accumulation effects of 126
Carica papaya, papain from 140
Casson body 120–1
Casson viscosity (K_c) 84
catabolite repression 36
  cAMP level and 46–7
cell culture, animal, for antibody production 3
cell division rate 6
cellulose, uses of 137
centrifugation, theory of 149–50
centrifuges
  performance of 149
  types of 149–51
cephaloridine, structure of 38
Cephalosporium acremonium 38
cephalosporin C, structure of 38
cephalosporins and derivatives 38
cephalothin, structure of 38
cephamycin C, structure of 38
champagne 12
cheese whey, production of in lactase reactor 196
chemical mutagens *see* mutagens
chemostat
  continuous-culture 23–5
  methane-limited 25
cholesterol oxidase, release of from *Nocardia* cells 148
chromosome rearrangements 49–51, 60
citric acid, production of 17–18
  using *Aspergillus foetidus* 88
Claviceps purpurea 8–9
clavulanic acid, structure of 30

cloned genes, expression of 68–9
cloning
  of DNA 61–73
  of cDNA for preproinsulin 74–5
coefficient of rigidity of fluids 119–20
coherent moving-bed system of fermentation 90
cohesive (sticky) end in cleaved DNA 62–3
complementary DNA (cDNA) *see* DNA
completely-mixed continuous system 91–3
co-mutation 49–51
concerted feedback (inhibition and repression) 56
conical fermenter 98–9
conjugation, bacterial 53–4
consistency coefficient of fluids 119–20
constitutive enzyme 44
contamination, prevention of in fermentation processes 86–7
continuous and batch processes, comparison of 91–3
continuous culture
  definition of terms used in 23–5
  fermentation by 23–5
continuous operation, engineering aspects of 90–1
continuous reactors (CSTR) for enzymes 184, 186–7
continuous stirred-tank fermenters 104–7
control and instrumentation in fermenters 101–4
cooling as scale-up problem 126–7
cooling jackets and coils 120–1
cooling systems in fermenters 111–12
corn steep liquor in penicillin production 30–1
corrosion, avoidance of in fermenters 99
Corynebacterium, amino acid production by 12–13
C. glutamicum 14–15
C. glycinophilum, L-serine production by 16–17
cost of enzyme cofactors 138–9
covalent binding, immobilization method and 171
crossing over, mitotic 51
CSTR, growth of *Candida lipolytica* in 25–7
cyanogen bromide, use of in cleavage 74–5
cysteine biosynthesis in *Penicillium chrysogenum* 32–3

deep-jet fermenter 108
'deep-shaft' process 131–3
deletion mutation 49–51
denitrification by anoxic process 90
design principles for fermenters 97–110

INDEX

detergents for cell disruption 148
diagnostic reagents 76–8
diaphragm valves in fermenters 99–100
diffusion coefficient (D_e)
  with immobilized enzymes 178–9
  relationship of to Thiele modulus 179
diffusion limitation, internal or external 178–9
dilatant fluids 120–1
dilution rate (D)
  in continuous culture
  optimization effect of on protein biosynthesis 30
disinfectants for sterilization 113–15
disruption techniques for cells
  chemical 148–9
  physical 145–8
dissolved oxygen (DO), control of 102
DNA
  chemical mutagens and 49–51
  cloning of 61–9
  cohesive (sticky) ends of 62–3
  complementary (cDNA) 67, 68, 72, 74
  cutting and joining techniques for 61–9
  errors in replication of 49–51
  fingerprinting using 72
  flush ends in 62–3
  hybridization technique for 67
  in genetic engineering 61–73
  ligase (T4) for 69
  mutations in 49–51
  recombinant 61–73
  palindromic recognition site in 62–3
  for preproinsulin 74
  probes using complementary 67, 72
  recognition and cleavage sites in 62–3
  repair mechanisms in 49–51
  replication point in 49–51
downstream processing, product recovery by 197–8
'down' time 91–3
dyno-mill for large-scale cell disruption 146

ecological advantage of antibiotics 7–8
economic factors in fermentation 83–4
Eco R1 restriction endonuclease 62–3
Eco R1 site 73
effectiveness factor
  of immobilized enzyme 174
  'operational' 173–4
  relationship of to Thiele modulus 177
  'stationary' 173–7
Einstein equation, viscosity relationship and 120
Embden–Meyerhof–Parnas (EMP, glycolysis) pathway 17–18, 40–1

encapsulation, immobilization method and 173–4
Endomycopsis fibuliger in Symba process 27
endonucleases, restriction, in genetic engineering 61–73
endotoxins, avoidance of risk from in vaccines 76–8
entrapment, immobilization method and 172–3
enzyme activity, regeneration of 194–5
enzyme catalysts, various forms of 165–8
enzyme, constitutive 44
enzyme engineering, stabilization by 137
enzyme, immobilized, factors modifying activity of 177–8
enzyme inactivation 144
enzyme induction (maltase and lactase) 2
enzyme isolation by concentration techniques 153–7
enzyme kinetics
  in batch reactor 190–2
  in continuous-flow reactor 190–2
  practical 190–201
  in tubular or plug-flow reactor 190–2
  under industrial conditions 174–81
enzyme loading
enzyme overproduction, genetic methods of 54–5
enzyme production, increase in by gene cloning 69–70
enzyme purification, chromatographic 157–63
enzyme reactors
  constant productivity from 194
  efficiency comparison of 186–7
  electrochemical 190
  fluidized-bed 198–9
  material balance in 185
  performance assessment of 183–90
  plug-flow 183
  tubular 90
  types of 183
  ultrafiltration 189
enzymes
  advantage of use of 136
  antibiotic-committed 8–9
  in brewing 139–40
  in industry 137, 138–9
  lytic, for cell disruption 148
  for process improvement 137
  for product improvement 138–9
  and proteins, human 3
  redesign of 71, 80
  restriction, in genetic engineering 61–8
  sources of 143–5
  stabilization of 137
  thermostable 138

INDEX

enzymic rate of reaction
  effect of pH on 139
  effect of temperature on 139
enzymic reaction 'order' in industrial situation 175
equilibrium constant (K) in industrial enzymology 175–7
ergoline alkaloids 8–9
error-free mechanisms in DNA restoration 49–51
error-prone repair pathway 49–51
erythromycin 39–40
Escherichia coli
  energy charge in 10
  mutant of 45
ethanol as fermentation substrate 27
eukaryotic genes, introns in 68
evaporation, enzyme concentration by 156–7
explosion, oxygen as hazard in 88–9
exponential phase of growth 6–7
expression of cloned genes 68–9
extrachromosomal elements 49–51

feedback inhibition, aspartate kinase and 14–15
feedback mechanism
  hypothetical outline of 47
  L-phenylalanine biosynthesis and 16–17
fermentation 5
  acetic-acid (vinegar) 18–20
  air sterilization in 101
  continuous-culture 23–5
  economic factors in 83–4, 133–4
  gluconic-acid 18
  heat sterilization in 113
  itaconic-acid 18
  mixed-culture 25–6
  organic-acid 17–20
  principles of in amino acid production 16–17
  productivity of 94–6
  sterile or aseptic conditions in 86–7
  temperature control in 110–12
  waste-water treatment by 130
fermentation engineering
  notation used in 95
  in SCP production 131–3
fermentation processes 128–34
fermentation substrates, starchy and sugary 11–12
fermenters
  agitation 88–9
  air-lift 17–18, 107–10, 131–3
  aspect ratios of 34
  baffles in 104–6
  biological environment in 86–7
  bubble-column 107–10
  chemical environment in 86–7
  in citric acid production 17–18
  conical 88–9
  control and instrumentation in 101–4
  cooling systems in 111–12
  deep-jet 108
  definition of 5
  design principles of 97–100
  dissolved oxygen control in 102
  foam-breaking in 103
  gas-lift 107–10
  mixing in 116–22
  oxygen supply in 122–6
  in penicillin manufacture 129–30
  physical environment in 87
  pumps in 100
  in SCP production 131–3
  sparged-tank 107–10
  sparging-in of air in 88–9
  stainless steel for 99
  steam in 98–100
  sterile audit of 101
  stirred-tank 104–7
  suspended-growth systems in 88–90
  types of 87–90
  unit size of 93
  valves in 99–100
fermenting organisms 12–13
fibroblast interferon 75–6
Fick's law of diffusion 173–7
filamentous fungi, fluid-type 119–20
films, microbial
filters for sterilization 115
filtration and filter aids in enzyme isolation 152–3
fingerprinting, DNA 72
flocculation and coagulation in enzyme isolation 152–3
flow behaviour index of fermentation media 120
flow pattern of liquid in fermenters 105–6
flow rate
  effect of on degree of enzymic conversion 191
  non-ideal, effect of on enzyme performance 192–4
  reactor, pressure drop caused by 197–8
fluidized beds 89–90
fluidized-bed enzyme reactor 188–9
fluids, visco-elastic 118–20
flush ends in cleaved DNA 62–3
foam-breaking in fermenters 103–4
foodstuffs, toxicity of 138–9
frameshift mutation 49–51
freeze-drying (lyophilization), enzyme concentration by 156

freezing and thawing, cell disruption by 147
freezing, enzyme concentration by 156
Frings acetator in vinegar production 22
Fusarium graminearum, growth of on carbohydrate-containing waste 18–20

β-galactosidase induction, blocking agent and 54–5
gas bubble or slug 107
gas transfer 88–9
gaseous cavitation in ultrasonication 147
gel chromatography, theory of 158–61
gene
　attenuator region of 45
　copies of 45
gene cloning of interferon 75–6
gene copy number 49–51
gene for insulin, chemical synthesis of 74–5
genes
　cloned, expression of 68–9
　introns in 68
genetic manipulation
　basic techniques of 61–9
　in-vivo 48–61
genetic engineering 44–82
　future prospects of 80
　products of 74–8
　in Pruteen production 30–1
　safety implications of 78–9
ginsenosides 42
gluconic acid, production of 18
glucose isomerase
　immobilized form of 137
　uses of 141–2
glucose oxidase from *Aspergillus niger* 18
glutamate dehydrogenase gene, transfer of 29–30
glutamate synthase (GOGAT) 29–30
glutamic acid bacteria 14
L-glutamine, production of 13
glutamine synthetase (GS) 29
glutaraldehyde as cross-linking reagent 172–3
glycolysis pathway 13
glycyrrhizin 42
5′-GMP, production of 21–2
gramicidin S, biosynthetic route to 9
gramicidin synthetase 8–10
gramicidins as secondary metabolites 7–9
growth
　exponential-phase 5–6
　lag-phase 5–7
　stationary phase 6–7
growth hormone *see* hormone
growth-limiting substrate (gls) in chemostat 23–5

Haemophilus influenzae, transformation in 53–4
Hansenula anomala, L-tryptophan production by 16–17
haploid spores, genetics of 51–2
haploidization frequency 51–2
heat sterilization in fermentation 113
heat transfer
　coefficient of 110–12
　in fermenters 110–12
hepatitis antigen 76–8
heterogeneous catalysis, immobilized enzymes in 175–7
heterokaryon, fungal 50–1
high-fructose syrups 167
*Hin*d III restriction endonuclease 62–3
homogenization of cells 145
homopolymer tailing in ligation to vectors 67
homoserine dehydrogenase 14–15
homoserine kinase 14–15
hormone
　genetic engineering of 70, 73–5
　human growth 70, 73–5
hosts (other than *E. coli*) 65
Hpa 1 restriction endonuclease 62–3
human genes 3
human growth hormone *see* hormone
human proteins, cell culture and 3
hydraulic residence time in fermenters 91–3
hydrostatic pressure in air-lift fermenters 110–12

immobilization methods for biocatalysts 171–4
immobilization support 171–5
　assessment of 173–4
immobilized bacteria in production of organic acids 173–4
immobilized biocatalysts 168–82
immobilized cells 168–82
immobilized enzyme (IME) 164–200 (*see also* enzymes)
　effectiveness factor in 174
5′-IMP, production of 21–2
impellers in fermenters 104–7
inactivation of enzymes 144
incremental operation 91–3
induction
　in eukaryotes 44–7
　of β-galactosidase 44–7
　of *lac* operon of *E. coli* 44–7
　of maltase and lactase 2
　in prokaryotes 44–7
industrial conditions, enzyme kinetics in 174–81

inoculum development in *Penicillium chrysogenum* 34–5
inoculum size and volume 5–6, 88
insertion mutation 49–51
insulin
  cloning of in yeast 75–6
  genetic engineering of 75–6
  glycoprotein nature of 75
  molecular weight of 75–6
  purification of using anti-interferon monoclonal antibodies 162–3
  sources of 75–6
introns in DNA 68
inversion in mutation 49–51
ion exchange chromatography 159–60
L-isoleucine production and α-aminobutyrate addition 16–17
itaconic acid, production of 17–18

$k_L \cdot a$ (mass transfer parameter) 122–3
  in fermenters 122–3
  independence of from DO concentration 126–7
  units of 84
K_m (Michaelis constant)
  of enzyme, definition of 139
  partitioning effects on 177–8
Krebs cycle
  citric acid production in 17–18
  itaconic acid production in 17–18

lac promoter, use of in genetic engineering 73
β-lactamase and penicillin resistance 55–6
lactase process 196
  use of 141–2
lager production, *Saccharomyces carlsbergensis* and 12–13
lagoons, fermentation of waste waters in 88–9
lag phase of growth 5–6, 39–40
lambda phage from *E. coli* 64–5
large-scale operations, problems of 83–5
lethal mutations 49–51
leukocyte interferon 75–6
ligase, DNA, in genetic engineering 61–75
ligating enzymes for DNA 61–9
liquid shear 146
lyophilization (freeze-drying) in enzyme concentration 156
L-lysine, regulation of production of 14–15
lysogeny in phage 69
lysozyme 148
lytic enzymes 148

macrolide antibiotics 39
macrotetrolides 7–8

malt vinegar 18–20
malted barley 11–12
manipulation of genes *in vivo* 48–61
Manton–Gaulin homogenizer in yeast disruption 146
mashing
  of cereals 12
  process details of 12
mass transfer of oxygen 122
maximum specific growth rate (μ_{max}) 6–7
Mbo 1 restriction endonuclease 62–3
meat, tenderization of 140–2
medical research, benefits to from genetic engineering 73–8
membrane polarization in ultrafiltration membranes 156
messenger RNA 68
  for preproinsulin 74–5
methane in SCP production 27
methane-utilizing microorganisms 25–30
methanol, utilization and mass balance in 27–30
Methylophilus methylotrophus, growth of on methanol 27–30
Michaelis constant (K_m) of enzyme 139
Michaelis–Menten equation at low [S] 175
Mickle shaker for cell disruption 146
microbial cells as enzyme source 145
microbial community in methane-limited chemostat 25–7
microbial films 88
  polysaccharides in 19–20
microprocessors in fermentation processes 93, 101–4
mixed-culture fermentation 25–7
mixing in fermenters 116–22
monoclonal antibody
  affinity chromatography of 162–3
  production of in animal cell culture 3
Monod equation 7
Monascus purpureus 11
Mucor rouxii 11
multi-copy plasmids 64–5
multiple impellers in fermenters 105–6
mutagenesis, site-directed 71
mutagens, chemical 49–51
mutation, causes of 49–51

Nernst–Planck layer in diffusion limitation 178–9
Newtonian fluids in fermentation 118–20
nitrocellular disc replica plating technique 67
non-Newtonian fluids in fermentation 118–20
nonsense (stop) codon 73
notation for fermentation engineering 84

INDEX

novobiocin production, viscosity and 125–6
nucleic acids
  content of in SCP 25
  removal of 152
nystatin, production of in *Streptomyces noursei* 165

Occam's razor 64, 133–4
olivanic acids 40–1
optimum biocatalyst loading 178–9
organic acids, fermentation of 17–18
organic solvents in enzyme precipitation 153–4
ornithine, production of by *Corynebacterium glutamicum* 55–6
osmotic shock for cell disruption 126–7
overproduction of enzymes 54–5
oxygen
  as explosive hazard 88–9
  mass transfer of in fermentation 122–6
oxygen transfer rate (OTR), units of 84

palindromic DNA, recognition site for 62–3
papain, uses of 137–41
paper industry, uses of amylases in 139–40
penicillin
  general structures of 30–1
  overproduction of 56
  production of 30–1, 33
  strain development and 56–61
penicillin acylase, use of in immobilized form 137, 186–8
penicillin manufacture 36–8
  fermenters used in 129–30
Penicillium chrysogenum 30–1
  protoplast fusion in 53–4
  strain development of 56–61
P. roquefortii, protoplast fusion in 53–4
pectinase, uses of 137
Pekilo process 27
peptide antibiotic, precursors of 8–19
permeability, changes in bacteria 13–14
phage induction 69
pharmaceutical industry, batch process products for 92–6
phenylacetic acid as penicillin precursor 30–1
plant cell culture, products of 41–2
plant cells as enzyme sources 144–5
plasmid 49–51
  multicopy 64
plasmid vectors 64
plastic viscosity of fluids 118–20
plug flow
  in enzyme reactors 192–3
  yeast disruption in 146
plug-flow system, definition of 90–1

plug-flow enzyme reactors 187–8
polyethylene glycol (PEG)
  to precipitate enzymes 154–5
  in protoplast fusion 53–4
polysaccharides of microbial origin 19–20
post-translational modification of proteins *see* proteins
power input of fermenter 116–17
'power-law' fluids in fermenters 122
power number (N_p)
  definition of 116–17
  formula for 84
  relationship of to turbulence 116–17
preproinsulin, mRNA for 74–5
pressure-cycle fermenter for SCP production 107–10, 131–3
pressure drop, effect of flow rate on 197–8
primary metabolites 10
  definition of 5–6
  overproduction of 55–6
principles of fermentation in amino acid production 16–17
probes, cDNA 67, 72
process comparison, batch v. continuous 91–3
process improvement using enzymes 138–40
product recovery by downstream processing 197–8
production rate biomass 6–7
productivity ($D\bar{x}$)
  in continuous culture 23–5
  in fermenters 94–6
  variation of with dilution rate 23–5
L-proline, production of 14
  regulation of 14
Propionibacterium denitrificans in vitamin B_{12} production 21–2
protease, uses of 137
protein break in brewing 12
proteins
  and enzymes, human 3
  optimization of biosynthesis of 30
  post-translational modification of 47
  processing of 74–5
  secretion of 74–6
protoplast fusion 53–4
Pruteen (SCP) 30
  dimensions of fermenter for 132
  gene cloning details of 78
pseudoplastic fluids in fermentation media 119–20
Pst 1 restriction endonuclease 62–3
pumps
  in fermenters 100
  role of 108
pyrimidine dimers in UV mutations 49–51